Holt Mathematics

Chapter 9 Resource Book

HOLT, RINEHART AND WINSTON

A Harcourt Education Company

Orlando • **Austin** • New York • San Diego • London

ISBN 0-03-078306-2

1 2 3 4 5   170   09 08 07 06

CONTENTS

Holt Mathematics

Date ____________

Dear Family,

In this chapter, your child will learn how to find the perimeter and area of geometric figures, and the circumference and area of circles. Your child will also learn about powers and roots and how to solve problems using the Pythagorean Theorem. Your child will apply the concepts learned in this chapter to problems involving various contexts, such as art, geography, sports, and architecture.

The **perimeter** of a figure is the distance around a figure. Your child will learn that to find the perimeter of a polygon, you can add the side lengths.

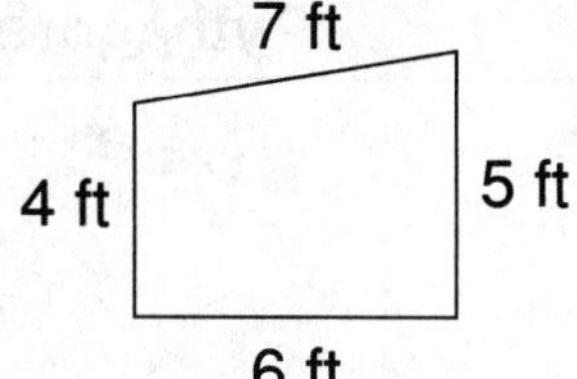

$P = 4 + 7 + 5 + 6 = 22$ ft

The **circumference** of a circle is the distance around a circle. To find circumference, your child will use the formula, $C = 2\pi r$.

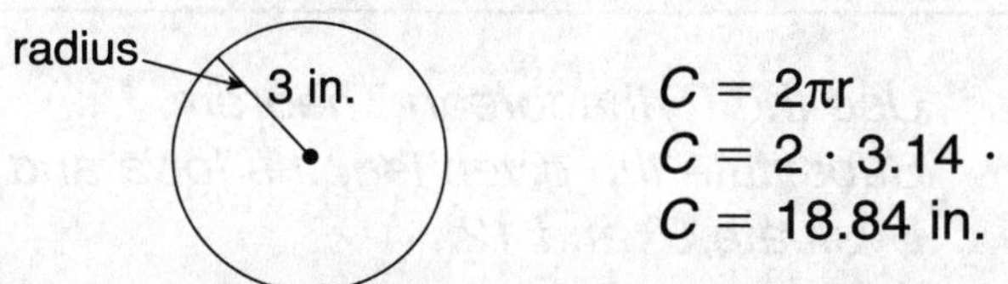

$C = 2\pi r$
$C = 2 \cdot 3.14 \cdot 3$
$C = 18.84$ in.

The **area** of a figure is the number of square units needed to cover it. Your child will use formulas to find the areas of various figures.

	Figure	Area Formula
Rectangle or Parallelogram		$A = lw$ or $A = bh$
Triangle		$A = \frac{1}{2}bh$
Trapezoid		$A = \frac{1}{2}h(b_1 + b_2)$
Circle		$A = \pi r^2$

1

Holt Mathematics

Your child will find the area of irregular figures, such as the one shown below.

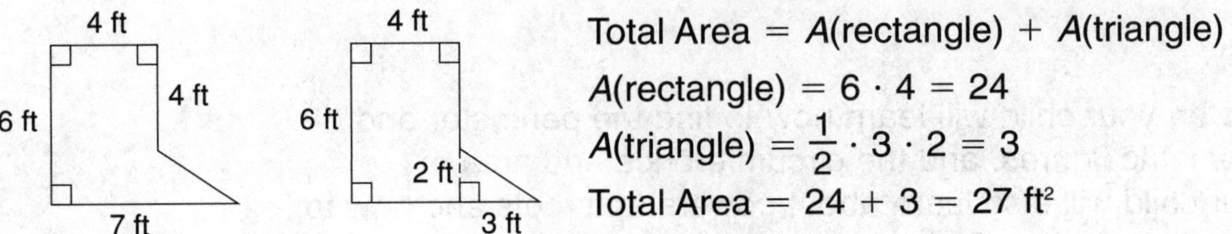

Total Area = A(rectangle) + A(triangle)

A(rectangle) = $6 \cdot 4 = 24$

A(triangle) = $\frac{1}{2} \cdot 3 \cdot 2 = 3$

Total Area = $24 + 3 = 27$ ft^2

In order to find the measure of a side of a right triangle, your child will learn how to use the Pythagorean Theorem.

Pythagorean Theorem		
In a right triangle, the sum of the squares of the lengths of the legs is equal to the square of the length of the hypotenuse.	$a^2 + b^2 = c^2$	5 cm, c, 12 cm

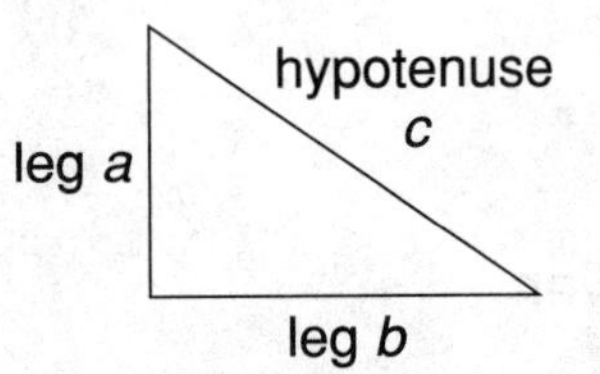

$a^2 + b^2 = c^2$	Use the Pythagorean Theorem.
$5^2 + 12^2 = c^2$	Substitute the given lengths for a and b.
$25 + 144 = c^2$	Evaluate 5^2 and 12^2.
$169 = c^2$	Add.
$13 = c$	Take the square root on both sides of the equation.

For additional resources, visit go.hrw.com and enter the keyword MS7 Parent.

Holt Mathematics

LESSON 9-1 Practice A
Accuracy and Precision

Choose the more precise measurement in each pair.

1. 5 ft, 53 in.   **2.** 10 qt, 44 c   **3.** 853 g, 1 kg

___________   ___________   ___________

Determine the number of significant digits in each measurement. Choose the letter for the best answer.

4. 3,000
- **A** 1
- **B** 2
- **C** 3
- **D** 4

5. 1.50
- **F** 1
- **G** 2
- **H** 3
- **J** 4

6. 3.405
- **A** 1
- **B** 2
- **C** 3
- **D** 4

7. 0.0006
- **F** 1
- **G** 4
- **H** 5
- **J** 6

8. 201
- **A** 2
- **B** 3
- **C** 4
- **D** 5

9. 50.020
- **F** 2
- **G** 3
- **H** 4
- **J** 5

Calculate. Find each answer with the correct number of significant digits in the box. Each answer can be used only once.

13.6	4.8	30	5	66	28	7.6	8	28.2	66.4	4.77	14

10. $19 - 5.4 =$ _______   **11.** $23.6 + 6 =$ _______   **12.** $11.8 - 7 =$ _______

13. $12.3 \cdot 5.4 =$ _______   **14.** $6.0 \cdot 4.7 =$ _______   **15.** $30.5 \div 4.0 =$ _______

16. $0.329 \cdot 14.49 =$ _______   **17.** $58.3 + 8.12 =$ _______   **18.** $59.04 \div 2.09 =$ _______

19. $32.6 - 27.83 =$ _______   **20.** $9 - 1.4 =$ _______   **21.** $12.9 + 0.71 =$ _______

Holt Mathematics

LESSON 9-1 Practice B
Accuracy and Precision

Choose the more precise measurement in each pair.

1. 2 tons, 3,700 lb

2. 4 weeks, 27 days

3. 3.5 m, 3.03 m

____

____

____

4. 3 ft, 32 in.

5. 4.6 mL, 2.8 L

6. 15.8 km, 15 km

____

____

____

Determine the number of significant digits in each measurement.

7. 5.801 _______

8. 0.06 _______

9. 75,000 _______

10. 0.00007 _______

11. 100,000,000 _______

12. 300.080 _______

13. 9.007 _______

14. 0.840 _______

15. 0.0050 _______

Calculate. Use the correct number of significant digits in each answer.

16. $21 - 8.6 =$ _______

17. $47.6 + 8 =$ _______

18. $9.8 - 3 =$ _______

19. $31.3 - 24.78 =$ _______

20. $9.63 + 3.4 =$ _______

21. $15.7 + 0.82 =$ _______

22. $0.54 + 0.104 =$ _______

23. $102 - 2.77 =$ _______

24. $62 + 0.319 =$ _______

25. $52.7 \cdot 2.3 =$ _______

26. $8.0 \cdot 1.7 =$ _______

27. $20.5 \div 6.0 =$ _______

28. $23.9 \cdot 14.4 =$ _______

29. $19.2 \div 0.03 =$ _______

30. $1{,}240 \div 4.025 =$ _______

31. $0.18 \cdot 6.2 =$ _______

32. $95 \div 32 =$ _______

33. $74.3 \cdot 0.22 =$ _______

Holt Mathematics

LESSON 9-1 Practice C
Accuracy and Precision

Choose the more precise measurement in each pair.

1. 6 L, 5,320 mL

2. 27 oz, 2 lb

3. 400 min, 6.5 hr

4. 5 ft, 60 in.

5. 8.7 km, 870 m

6. 8 cm, 2.5 cm

Determine the number of significant digits in each measurement.

7. 2.0801 _____

8. 0.03 _____

9. 10,500 _____

10. 0.00307 _____

11. 900,000,000 _____

12. 302.80 _____

13. 7.007 _____

14. 0.840 _____

15. 0.080 _____

Calculate. Use the correct number of significant digits in each answer.

16. $170 - 6.8 =$ _____

17. $37.6 + 8 =$ _____

18. $2.8 \div 0.3 =$ _____

19. $51.37 - 24.7 =$ _____

20. $6.89 + 7.3 =$ _____

21. $95.7 + 9.893 =$ _____

22. $72.5 \cdot 3.2 =$ _____

23. $2,600 \cdot 4.71 =$ _____

24. $25 \div 7.04 =$ _____

Which unit is more precise?

25. quart, gallon _____________

26. ounce, pound _____________

27. pint, cup _____________

28. day, month _____________

29. gram, kilogram _____________

30. inch, foot _____________

Holt Mathematics

LESSON 9-1 Reteach
Accuracy and Precision

A measurement is more **precise** than another measurement if its unit of measure is smaller.

---

1. Which measurement is more precise, 14 cm or 140 mm?

   a. 14 cm means the measure is to the nearest ___________.

   b. 140 mm means the measure is to the nearest ___________.

   c. Since a millimeter is a smaller measurement, ___________ is more precise.

2. Which measurement is more precise, 5 ft or 50.1 ft?

   a. 5 ft means the measure is to the nearest ___________.

   b. 50.1 ft means the measure is to the nearest tenth of a ___________.

   c. Since a tenth of a foot is a smaller measurement, ___________ is more precise.

Choose the more precise measurement in each pair.

3. 7.5 m or 75 cm   **4.** 11.0 in. or 11 in.   **5.** 8.4 lb or 8 oz

_________   _________   _________

You can find the number of **significant digits,** or all the digits that are known to be exact, by dividing a measurement by its smallest place value. The number of digits in the quotient is the number of significant digits.

Measurement	Smallest Place Value	Measurement ÷ Smallest Place Value	Number of Significant Digits
14 ft	1	14 ÷ 1 = 14	2
0.043 in.	0.001	0.043 ÷ 0.001 = 43	2
50.1 m	0.1	50.1 ÷ 0.1 = 501	3

Zeros to the left of a decimal point are not significant if there are no digits to the right of the decimal point.

   910 contains 2 significant digits: 9 and 1.

Zeros after digits to the right of a decimal point are significant.

   64.9500 contains 6 significant digits: 6, 4, 9, 5, and the two zeros.

Determine the number of significant digits.

6. 41.25 _______   **7.** 30.6 _______   **8.** 0.085 _______

9. 1,207,000 _______   **10.** 38.600 _______   **11.** 4.00020 _______

Holt Mathematics

Challenge
The Significance of a Riddle

Write a solution for each of the following riddles. Your answer must include all of the information in the riddle, but all of the information about each number is not always given in the riddle.

1. I am a number with *only* the digits 1, 2, and 3. I have four significant digits and no decimal places. What number could I be?

____

2. I am a number with *only* the digits 2, 4, and 0. I have one decimal place and four significant digits. What number could I be?

____

3. I am a number with *only* the digits 6 and 0. I have three significant digits and four decimal places. What number could I be?

____

4. I am the greatest number with *only* the digits 1, 2, and 3. I have five significant digits. What number am I?

____

5. I have no significant digits. What number am I?

____

6. What number could I be with two significant digits and five decimal places?

____

7. I am the least number with *only* the digits 1, 2, and 3, with five significant digits and six decimal places. What number am I?

____

8. I am a number with two significant digits and four decimal places. I begin and end with 0. What number could I be?

____

9. I am a number between 155 and 1,555. I have five significant digits and two decimal places. What number could I be?

____

10. I am a number between 0 and 2 with *only* the digits 1 and 4. I have five significant digits and four decimal places. What number could I be?

____

11. I have four digits, but only three are significant. I have three decimal places. What number could I be?

____

12. I have six digits, but only two are significant. I have no decimal places. What number could I be?

____

Holt Mathematics

Problem Solving

Accuracy and Precision

Write the correct answer.

1. Normal rainfall in Hilo, Hawaii, is 2.36 feet per year. Yearly rainfall in Honolulu, Hawaii, is 7.77 inches. Which measurement is more precise? Explain.

____

____

____

2. The Seismosaurus was about 5.5 meters high. The Troodon, considered the smartest dinosaur, was only about 1.75 meters high. Write the difference in height of the two dinosaurs using the correct number of significant digits.

____

3. Esther drives a total of 120 miles to and from work each day. She works 5 days per week. How many significant digits are there in the number of miles Esther drives to and from work each week?

____

4. Big Al weighs 256.8 pounds. His brother, Little Lou, weighs 125 pounds. What is their combined weight? Express your answer with the appropriate number of significant digits.

____

Choose the letter of the correct answer.

5. A square picture frame measures 50.1 centimeters on each side. How many significant digits are there in the perimeter of the frame?

A 2   **C** 4

B 3   **D** 1

6. Four students each measured his or her own height. The measures below show each student's results. Which is the most precise measure?

F 5 ft   **H** 62 in.

G 5.5 ft   **J** 64.5 in.

7. Before the year 2000, Harvard University had 12,877,360 books in its library. How many significant digits are there in the number of books in the Harvard Library?

A 5   **C** 7

B 6   **D** 8

8. In 1996, Tokyo Disneyland had an estimated 16.98 million visitors. Disneyland in Anaheim, California, had 15 million visitors. What is the estimated total number of visitors for both parks using the correct number of significant digits?

F 32 million   **H** 31.9 million

G 31.98 million   **J** 31 million

Holt Mathematics

Reading Strategies
Analyze Information

Precise means as close or accurate as possible. No measurement is absolutely exact. It is important when measuring to be as precise or exact as possible.
You can choose the more precise measurement.

The width of your little finger:
0.5 inch or 0.46 inch

Since 0.46 has a smaller place value than 0.5, it is the more precise measurement.

The width of a computer screen:
1 foot or 13 inches

Since inches are smaller than feet, 13 inches is the more precise measurement.

Complete each problem.

1. What does the word *precise* mean?

____

____

2. Which measurement is more precise, inches or feet? Explain.

____

____

3. Why is 0.46 inch more precise than 0.5 inch?

____

____

Circle the more precise measurement in each pair below.

4. $3\frac{1}{2}$ feet $3\frac{1}{3}$ feet

5. 34 mm   3 cm

6. 7 feet   2 yards

7. $\frac{1}{2}$ hour   35 minutes

8. 23 ounces   2 pounds

9. 1 hour   53 minutes

10. 27 days   1 month

Holt Mathematics

<table>
<tr><td>LESSON
9-1</td><td></td></tr>
</table>

Puzzles, Twisters & Teasers
Those Darn Details!

Find and circle the words below in the word search. Words may be horizontal or vertical or backwards. Find a word in the word search to solve the riddle. Circle it and write it on the line.

precision	measurement	determine	level	accuracy
digit	significant	exact	detail	accurate

```
A U P R E C I S I O N H M E
C M E A S U R E M E N T A N
C B Y G H D I G I T I O R I
U P L O K E V F E R G B T M
R B G T R T O I P L K Y I R
A C C U R A C Y A S D F A E
T Q W E R I T Y U H N I N T
E X A C T L E V E L Z X C E
Y S I G N I F I C A N T A D
```

What's an alien's favorite candy? _________________ - mallows

Holt Mathematics

Name ____ Date __________ Class ____________

Practice A
Perimeter and Circumference

Find the perimeter of each polygon.

1.
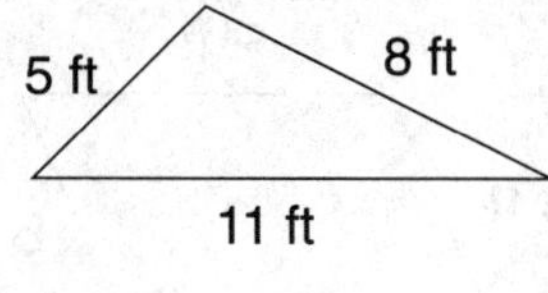

____

2.
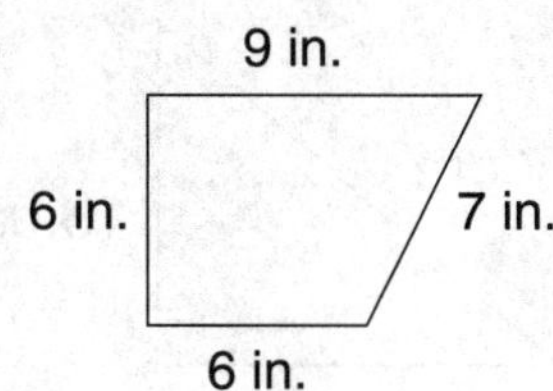

____

3.
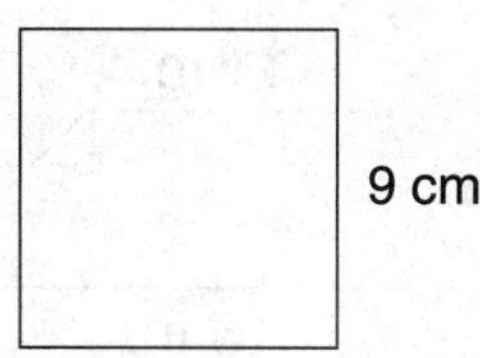

____

Find the perimeter of each rectangle.

4.
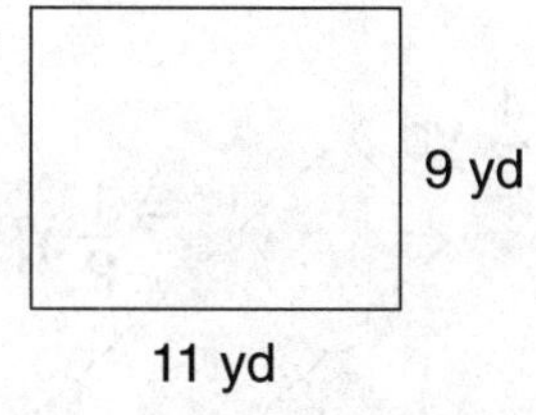

____

5.
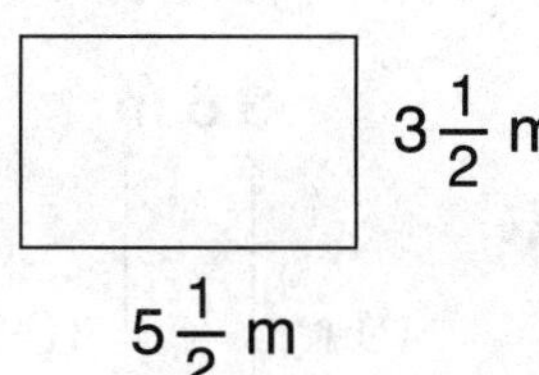

____

6.
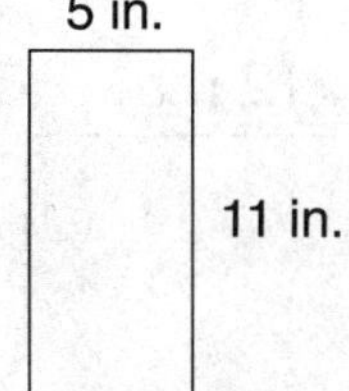

____

**Find the circumference of each circle to the nearest tenth.
Use 3.14 for π. Choose the letter for the best answer.**

7.
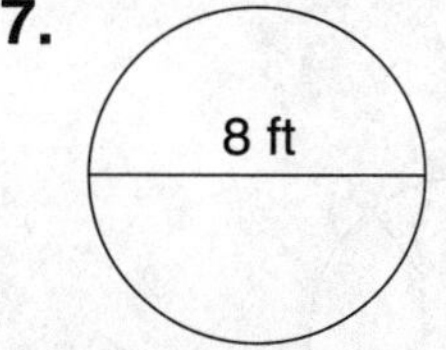

A 12.5 ft   C 25.1 ft

B 12.6 ft   D 25.2 ft

8.
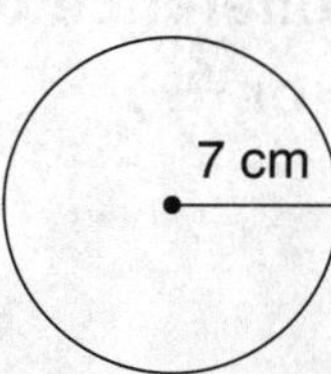

F 21.9 cm   H 43.9 cm

G 22.0 cm   J 44.0 cm

9.
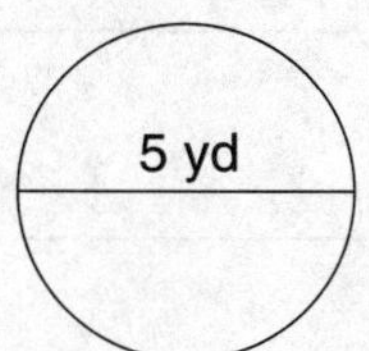

A 8.1 yd   C 25.2 yd

B 15.7 yd   D 31.4 yd

10.
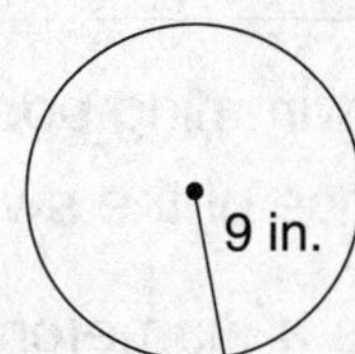

F 56.5 in.   H 28.2 in.

G 28.3 in.   J 14.1 in.

11. The diameter of a clock is 7 inches. What is the
circumference of the clock? Use $\frac{22}{7}$ for π. ____________________

Holt Mathematics

Practice B
Perimeter and Circumference

Find the perimeter of each polygon.

1. 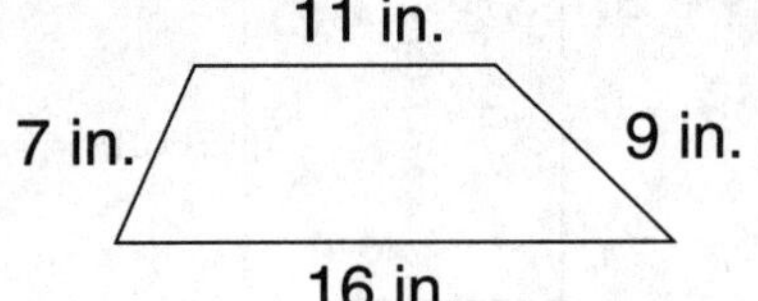

2. 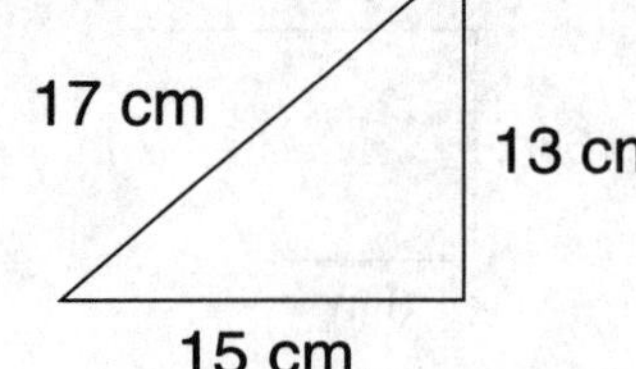

3. 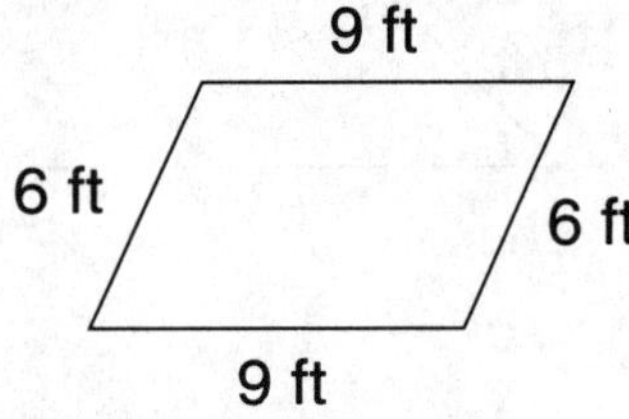

__________________ __________________ __________________

Find the perimeter of each rectangle.

4. 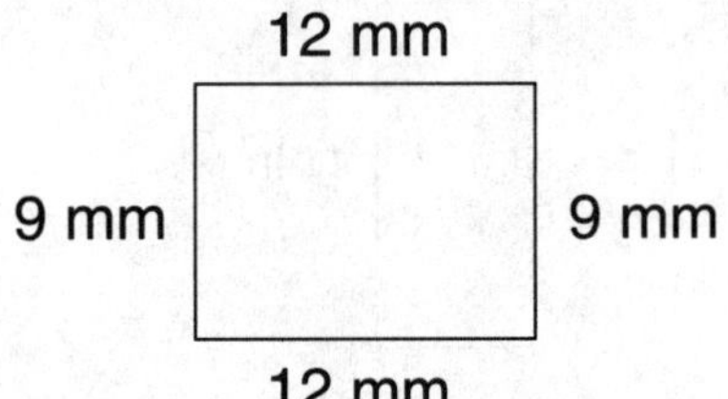

5. 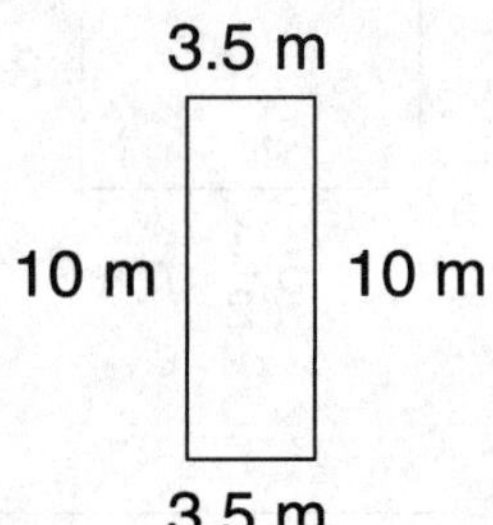

6. 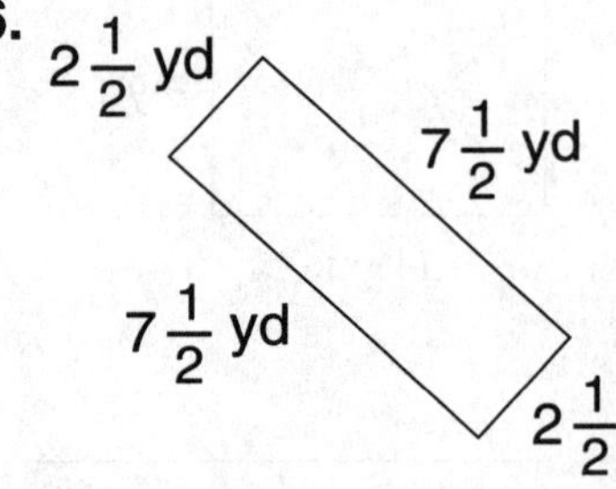

__________________ __________________ __________________

Find the circumference of each circle to the nearest tenth.
Use 3.14 for π or $\frac{22}{7}$.

7. 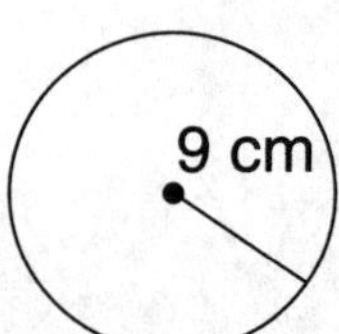

8. 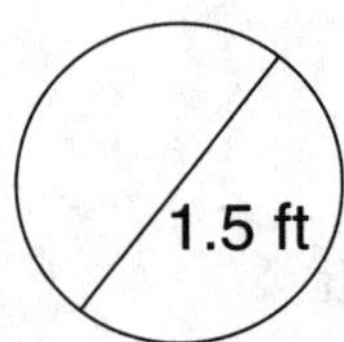

9.

__________________ __________________ __________________

10. A circular swimming pool is 21 feet in diameter. What is the
circumference of the swimming pool? Use $\frac{22}{7}$ for π.   __________________

11. A jar lid has a diameter of 42 millimeters. What is the
circumference of the lid? Use $\frac{22}{7}$ for π.   __________________

12. A frying pan has a radius of 14 centimeters. What is the
circumference of the frying pan? Use $\frac{22}{7}$ for π.   __________________

Holt Mathematics

LESSON 9-2 | Practice C
Perimeter and Circumference

Find the perimeter of each polygon.

1.
2.4 cm 2.3 cm
2.9 cm
3.1 cm 3.5 cm

____

2.
$7\frac{5}{16}$ in.
$5\frac{3}{16}$ in. $5\frac{3}{16}$ in.
$7\frac{5}{16}$ in.

____

3.
19.5 ft 7.3 ft
7.3 ft
7.3 ft
19.5 ft 7.3 ft

____

Find the perimeter of each rectangle.

4.
16 yd
49 yd

____

5.
17.2 m
8.9 m

____

6.
$8\frac{1}{2}$ in.
$3\frac{3}{4}$ in.

____

Find each missing measurement to the nearest tenth.
Use 3.14 for π or $\frac{22}{7}$.

7. $r = 14.5$ in.

$d =$ _______________

$C =$ _______________

8. $r =$ _______________

$d =$ _______________

$C = 35.6$ yd

9. $r =$ _______________

$d = 7.6$ cm

$C =$ _______________

10. $r =$ _______________

$d =$ _______________

$C = 11.3$ m

Solve.

11. In NCAA basketball rules, the basketball can have a maximum circumference of 30 inches. What is the maximum diameter of an NCAA basketball to the nearest hundredth? _______________

Holt Mathematics

LESSON 9-2 Reteach
Perimeter and Circumference

The **perimeter** of any figure is the distance around the figure. Think of *perimeter* as "going around the rim" of a figure. Find the perimeter of a figure by adding the measures of the sides.

Find the perimeter of each polygon.

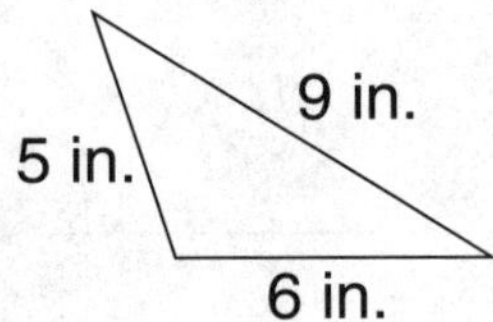

$P = 6 + 5 + 9$

$P = 20$

The perimeter of the triangle is 20 in.

$P = 7 + 11 + 10 + 8$

$P = 36$

The perimeter of the quadrilateral is 36 cm.

To find the perimeter of a rectangle, you can add the length and the width and multiply the sum by 2. The formula for the perimeter of a rectangle is $P = 2(\ell + w)$.

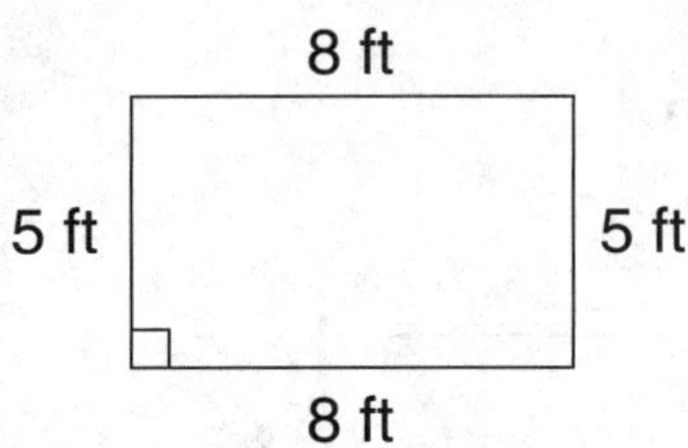

$P = 2(\ell + w)$

$P = 2(8 + 5)$

$P = 2(13)$

$P = 26$

The perimeter of the rectangle is 26 feet.

Find the perimeter of each polygon.

1.

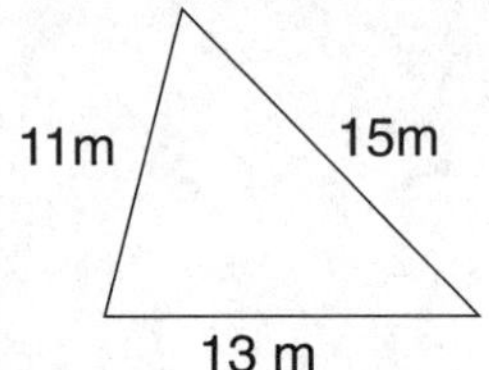

$P = $ ___ + ___ + ___

$P = $ ___ m

2.

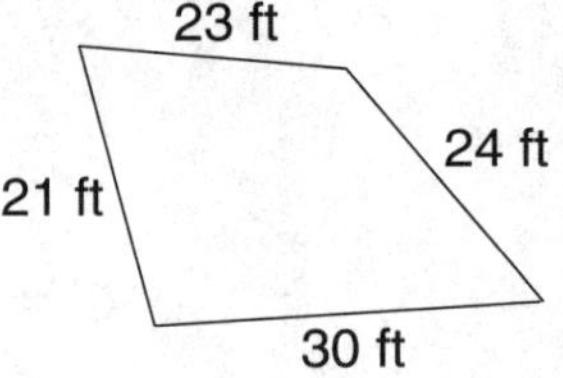

$P = $ ___ + ___

+ ___ + ___

$P = $ ___ ft

3.

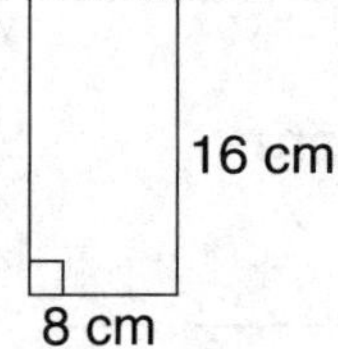

$P = 2($ ___ + ___ $)$

$P = $ ___ cm

Holt Mathematics

Reteach
Perimeter and Circumference (continued)

The distance around a circle is called the **circumference.** To find the circumference of a circle, you need to know the diameter or the radius of the circle.

> The ratio of the circumference of any circle to its diameter $\left(\frac{C}{d}\right)$ is always the same. This ratio is known as π (pi) and has a value of approximately 3.14.

To find the circumference C of a circle if you know the diameter d, multiply π times the diameter. $C = \pi \bullet d$, or $C \approx 3.14 \bullet d$.

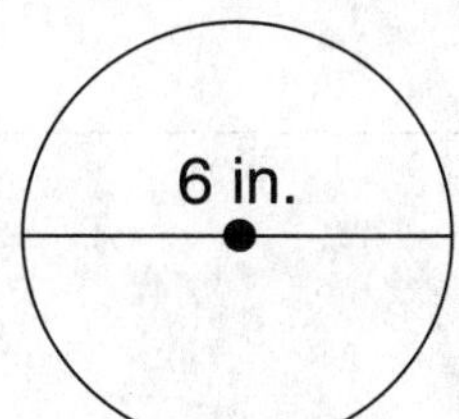

$C = \pi \bullet d$
$C \approx 3.14 \bullet d$
$C \approx 3.14 \bullet 6$
$C \approx 18.84$
The circumference is about 18.8 in. to the nearest tenth.

The diameter of a circle is twice as long as the radius r, or $d = 2r$. To find the circumference if you know the radius, replace d with $2r$ in the formula. $C = \pi \bullet d = \pi \bullet 2r$

Find the circumference given the diameter.

4. $d = 9$ cm

   $C = \pi \bullet d$

   $C \approx 3.14 \bullet$ _______

   $C \approx$ __________

   The circumference is _______ cm to the nearest tenth of a centimeter.

Find the circumference given the radius.

5. $r = 13$ in.

   $C = \pi \bullet 2r$

   $C \approx 3.14 \bullet (2 \bullet$ _______$)$

   $C \approx 3.14 \bullet$ _______

   $C \approx$ __________

   The circumference is _______ in. to the nearest tenth of an inch.

Find the circumference of each circle to the nearest tenth. Use 3.14 for π.

6.

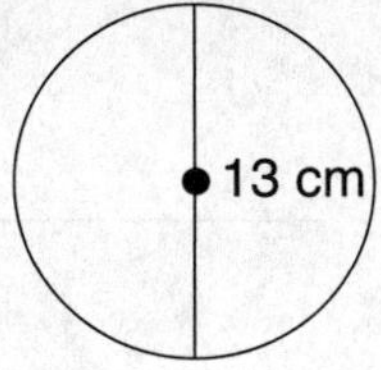

____

7.

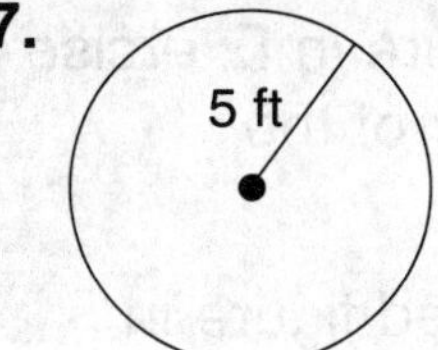

____

8.

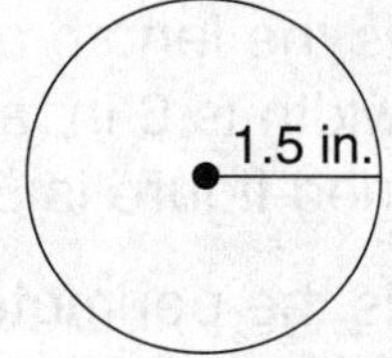

____

Holt Mathematics

LESSON 9-2 · Challenge
All-Around Formulas

Use the first figure in each row to write a formula for the perimeter of the combined figure next to it. Use your formulas in Exercises 5–8.

Original Figure	Combined Figure	Perimeter
1. A regular octagon: *m*		
2. An isosceles triangle: *s*, *b*		
3. A parallelogram: *w*, ℓ		
4. A semicircle: *d*		

5. What is the perimeter of the combined figure in Exercise 1 if $m = 4$ in.? ______________

6. What is the perimeter of the combined figure in Exercise 2 if $s = 4.5$ m and $b = 5.2$ m? ______________

7. What is the length of the original figure in Exercise 3 if the width is 6 in. and the perimeter of the combined figure is 56 in.? ______________

8. What is the perimeter of the combined figure in Exercise 4 if $d = 10$ cm? ______________

Holt Mathematics

Problem Solving
Perimeter and Circumference

Write the correct answer.

1. Mr. Marcos, the gym teacher, had the seventh graders run around the perimeter of the gym 3 times. The gym has a length of 34 feet and a width of 58 feet. What was the total distance the students ran?

2. The distance between bases on a baseball field is 90 feet. If 3 players hit home runs during a game and each runs around all 4 bases, what is the total distance the players run?

3. Basketball rims have a diameter of 18 inches. If you want to put a band around a basketball rim, how much material to the nearest tenth of an inch will you need?

4. A pizza cutter has a diameter of 2.5 inches. To cut a pizza in half, the cutter makes two complete revolutions. What is the diameter of the pizza?

5. A round stained-glass window has a circumference of 195 inches. What is the radius of the window to the nearest inch?

6. A planter full of pansies has a diameter of 14 inches. What is the circumference of the planter to the nearest inch?

Choose the letter of the correct answer.

7. A welcome mat on the front porch is a semicircle. The straight side of the mat is 36 inches. What is the perimeter of the mat?

   A 92.52 in.   C 56.52 in.

   B 64.26 in.   D 28.26 in.

8. The radius of the planet Jupiter is about 44,368 miles. What is the approximate circumference of Jupiter to the nearest mile?

   F 557,262 mi   H 139,316 mi

   G 278,631 mi   J 69,658 mi

9. Four square tables with sides of 48 inches each are placed end to end to form one big table. What is the perimeter of the table that is formed?

   A 192 in.   C 480 in.

   B 384 in.   D 768 in.

10. Three sides of the Great Pyramid at Giza, Egypt, each measure 756 feet in length to the nearest foot. If the perimeter of the pyramid is 3,023 feet, what is the length of the fourth side of the pyramid?

   F 754 ft   H 756 ft

   G 755 ft   J 757 ft

Holt Mathematics

<table><tr><td>**LESSON**
9-2</td><td>

Reading Strategies
Use A Graphic Organizer
</td></tr></table>

Perimeter is the distance around a polygon.

Circumference is the distance around a circle.

The chart below shows formulas for finding the perimeter of polygons and the circumference of circles.

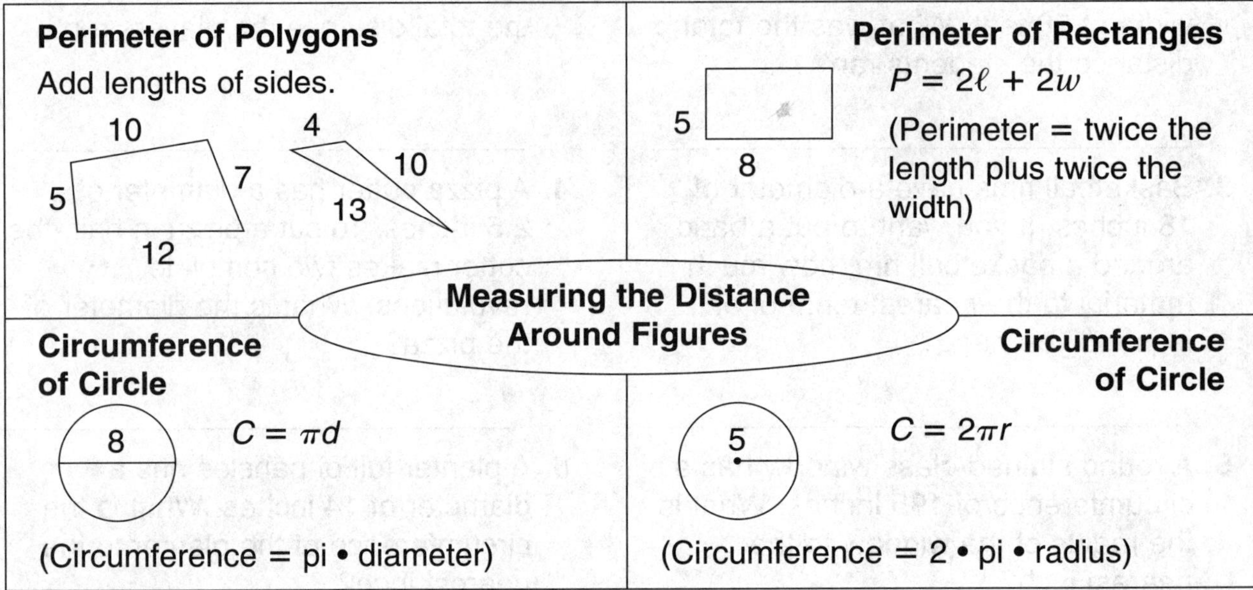

Use the information in the chart above to complete each exercise.

1. Write the lengths for the triangle shown above.

____

2. How would you find the perimeter of the triangle?

____

3. If you knew the radius of a circle, what formula would you use to find its circumference?

____

4. Write the formula for finding the perimeter of a rectangle.

____

5. If you knew the diameter of a circle, what formula would you use to find its circumference?

____

  Holt Mathematics

<table><tr><td>**LESSON**
9-2</td><td>

Puzzles, Twisters & Teasers
Around and Around!
</td></tr></table>

Find the perimeter or circumference of each figure. Round the circumference to the nearest tenth. Match your answers to the corresponding letter to solve the riddle.

1. __________ = **E**

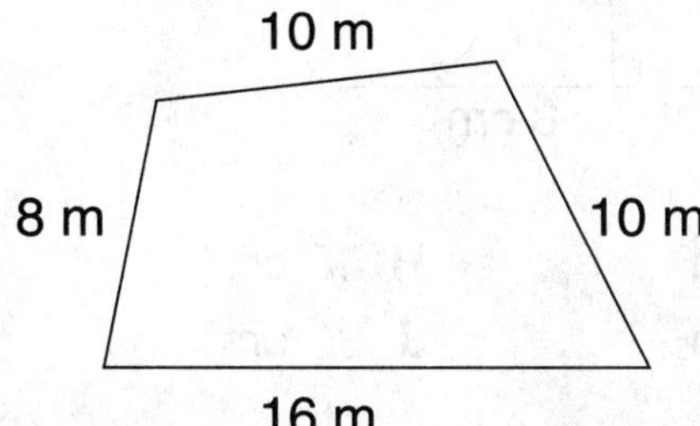

2. __________ = **V**

6.2 in.

3.7 in.

3. __________ = **L**

$3\frac{3}{4}$ ft

$1\frac{1}{4}$ ft

4. __________ = **I**

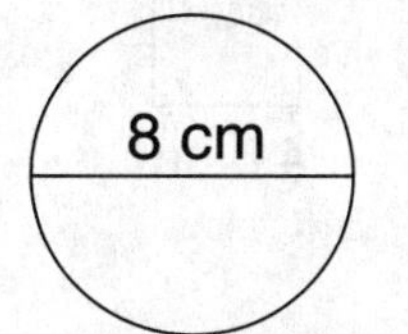

5. __________ = **C**

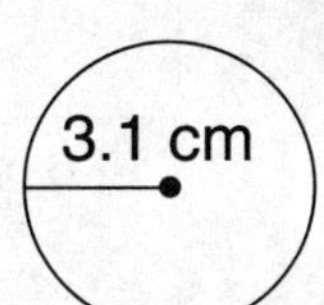

6. __________ = **O**

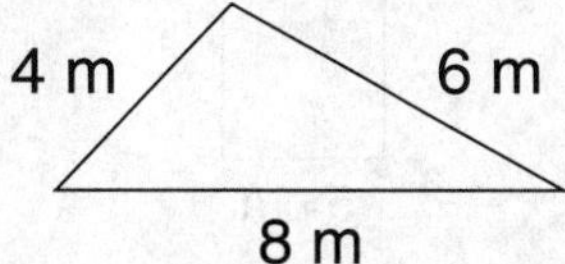

7. __________ = **R**

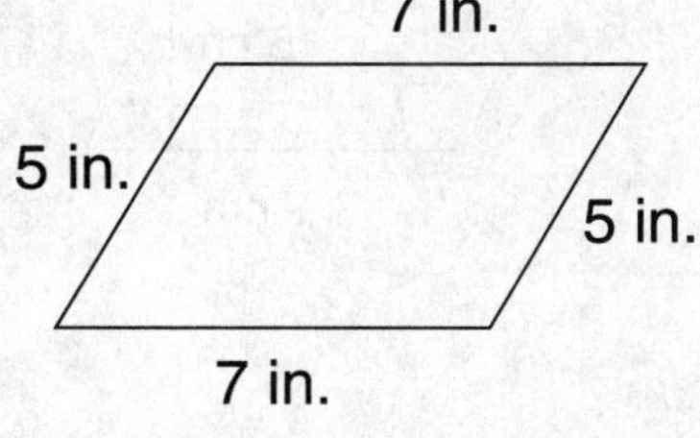

8. __________ = **S**

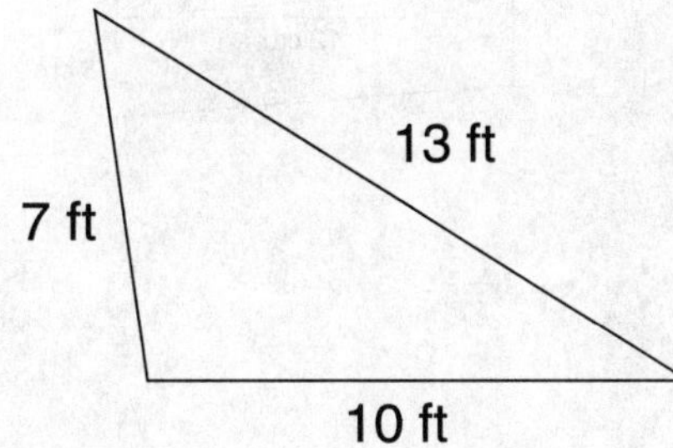

9. __________ = **U**

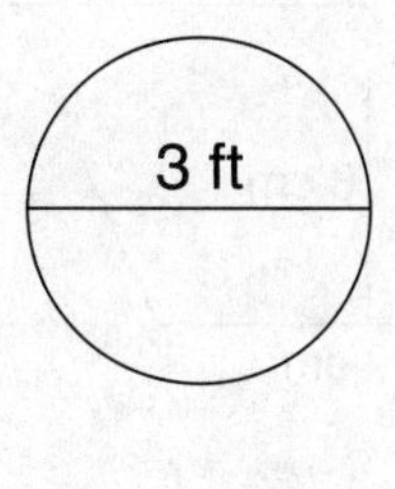

What's round and bad-tempered?

A __________ __________ __________ __________ __________ __________ __________
  19.8   25.1   19.5   25.1   18   9.4   30

__________ __________ __________ __________ __________ __________
  19.5   25.1   24   19.5   10   44

Holt Mathematics

Practice A
Area of Parallelograms

Find the area of each figure. Choose the letter for the best answer.

1.

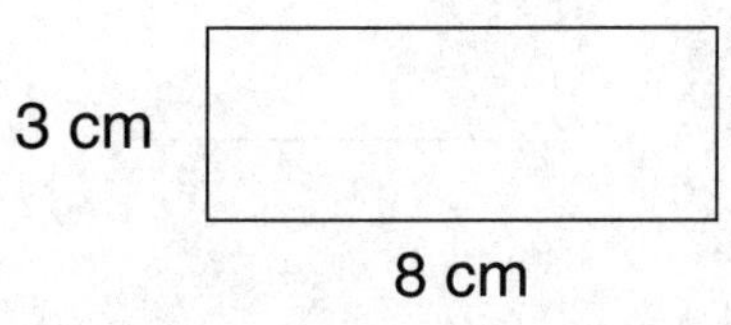

A 11 cm²   **C** 24 cm²

B 22 cm²   **D** 48 cm²

2.

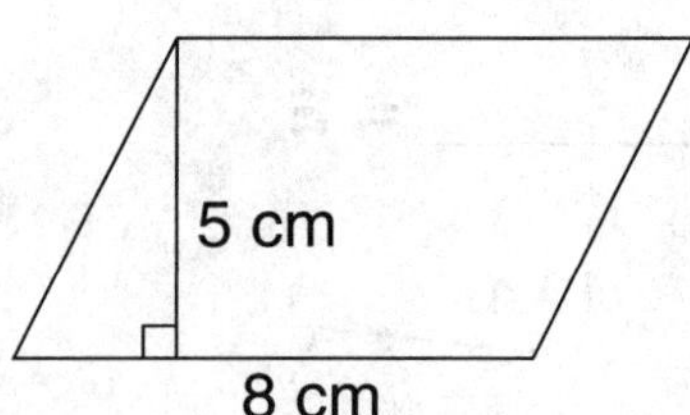

F 26 cm²   **H** 48 cm²

G 40 cm²   **J** 80 cm²

Find the area of each rectangle.

3.

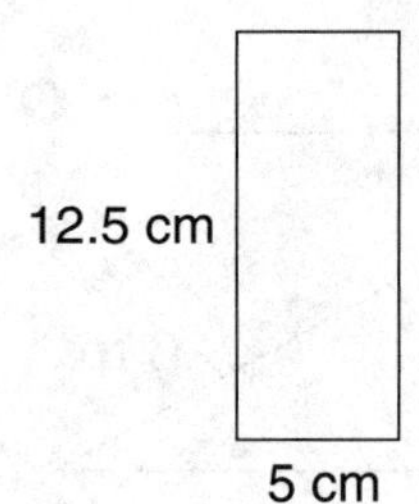

4.

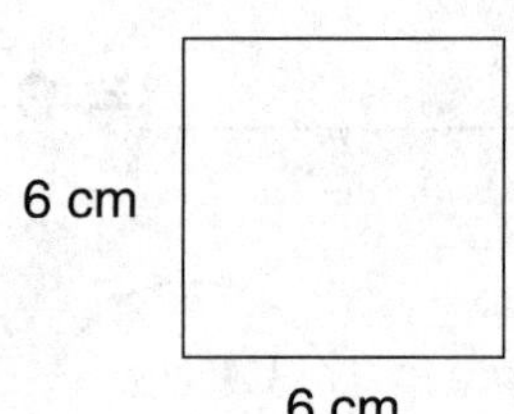

5.

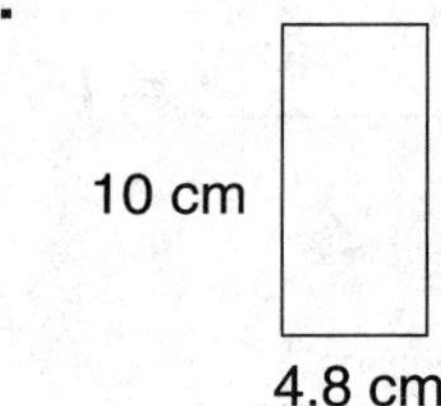

______________ ______________ ______________

Find the area of each parallelogram.

6.

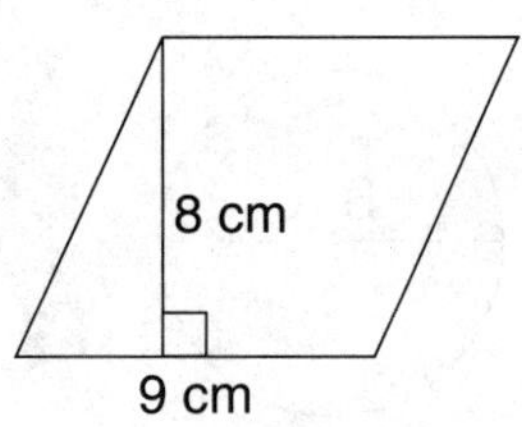

7.

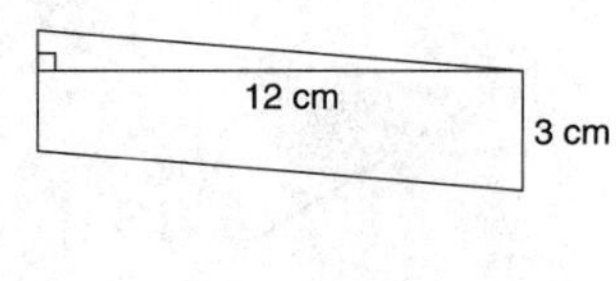

8.

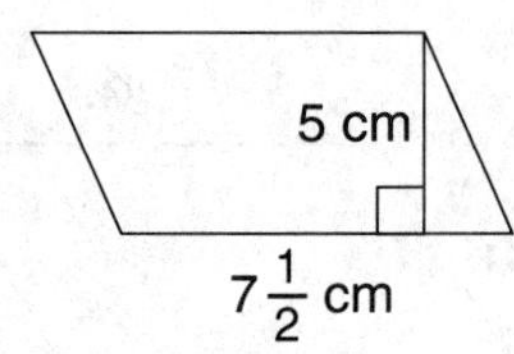

______________ ______________ ______________

9. Michelle wants to carpet her living room. The area of the living room is 192 ft². The length of the living room is 16 ft. What is the width of the living room? __________

10. Mustafa is tiling his bathroom. The section that needs to be tiled is 62 in. by 70 in. How many square inches of tile does he need? ______________

Holt Mathematics

Practice B
Area of Parallelograms

Find the area of each rectangle.

1. 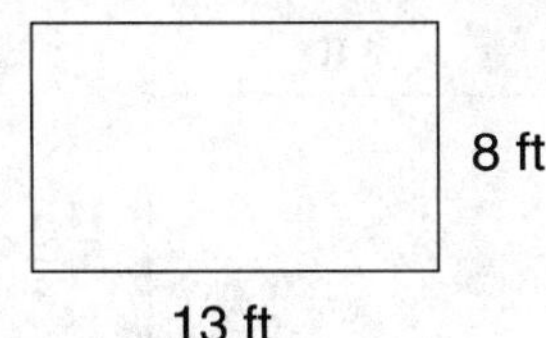

2. 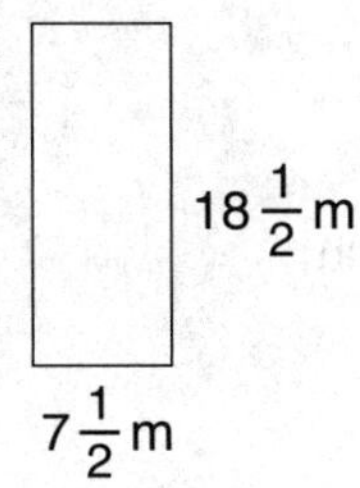

3. 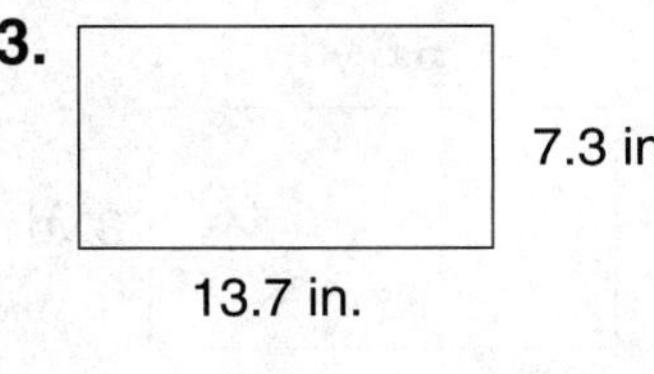

____

____

____

Find the area of each parallelogram.

4. 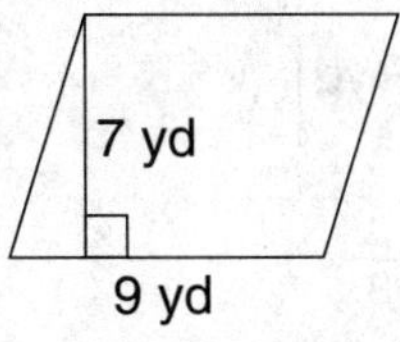

5. 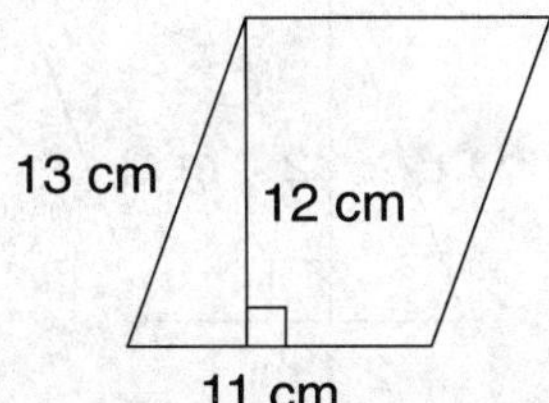

6. 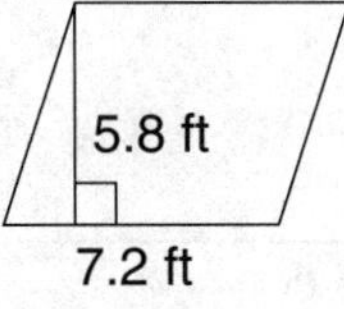

____

____

____

7. 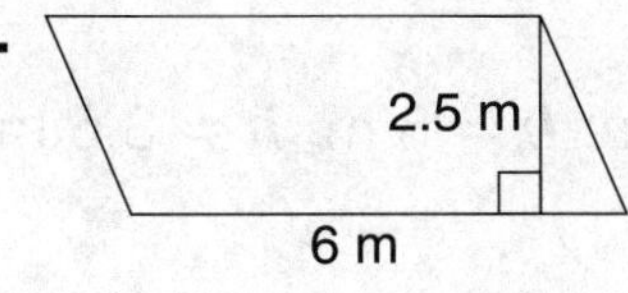

8. 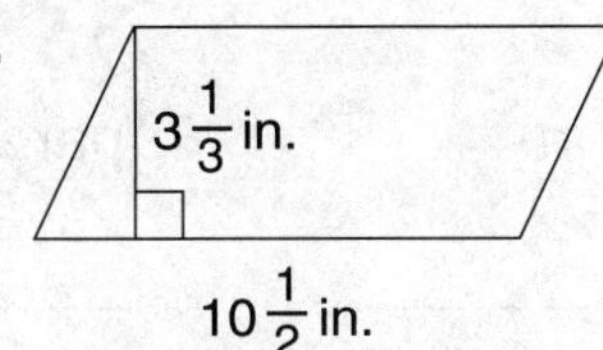

9. 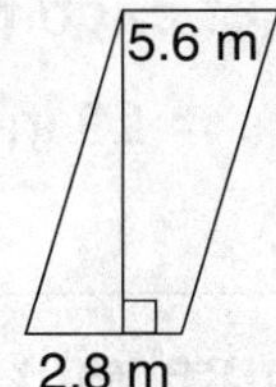

____

____

____

10. A dollar bill is 15.5 cm long and 6.5 cm wide. What is the area of a dollar bill?

____

11. A rectangular hallway has an area of 70 ft2. The width of the hallway is 4 feet. What is the length of the hallway?

____

Holt Mathematics

Practice C
Area of Parallelograms

Find the area of each rectangle.

1.
15.6 yd

8 yd

____

2. 2.5 in.

7.2 in.

____

3.
8 ft

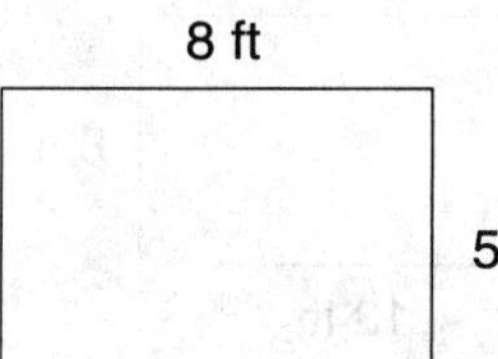
$5\frac{3}{4}$ ft

____

Find the area of each parallelogram.

4.
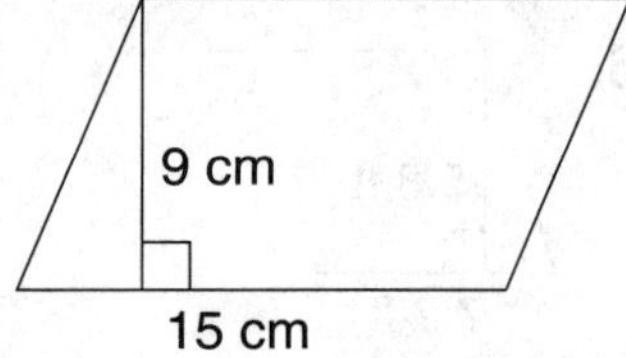
9 cm

15 cm

____

5.
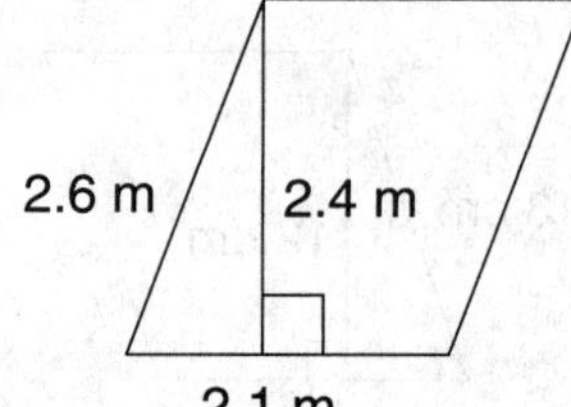
2.6 m   2.4 m

2.1 m

____

6.
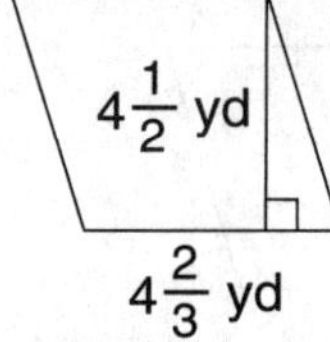
$4\frac{1}{2}$ yd

$4\frac{2}{3}$ yd

____

Find the area of each polygon.

7. rectangle: $\ell = 29$ in., $w = 15$ in.

____

8. parallelogram: $b = 8$ m, $h = 3.6$ m

____

Sketch the figure with the given vertices. Then find the area of the figure.

9. (3, −2), (8, −2), (3, 6), (8, 6)

____

10. (2, 0), (5, 3), (8, 0), (11, 3)

____

11. (−4,1), (−9, 1), (0, 7), (−5, 7)

____

12. (−1, −2), (4, −2), (−1, 5), (4, 5)

____

13. What is the length of the base of a parallelogram with an area of 36 cm^2 and a height of 4.5 cm?

____

14. A window is composed of 6 windowpanes. Each has a length of $10\frac{1}{2}$ in. and a width of $7\frac{1}{4}$ in. What is the total area of glass in the window?

____

Holt Mathematics

LESSON 9-3 Reteach
Area of Parallelograms

The **area** of a figure is the number of square units inside the figure.

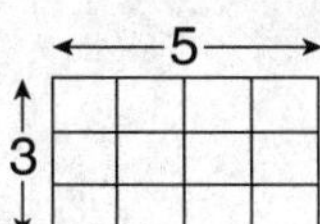

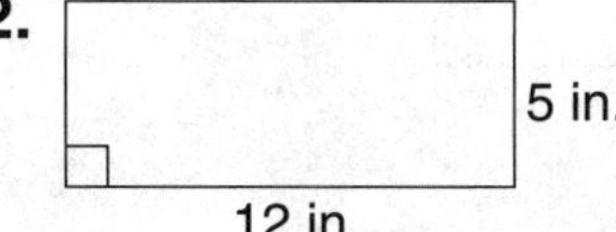

You can count the squares inside the rectangle. There are 15 square units within the rectangle. This is equal to 5 • 3.

To find the area of a rectangle, multiply the length (ℓ) times the width (w).
$$A = \ell \cdot w$$

Find the area of each rectangle.

1.

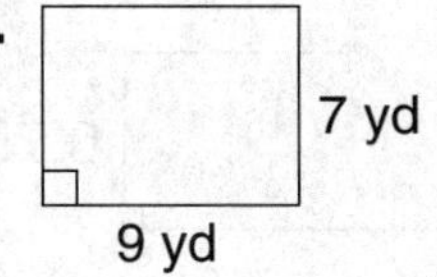
7 yd
9 yd

$A = l \cdot w$

$A = $ _____ • _____

$A = $ _____

The area is _____ yd^2.

2.

5 in.
12 in.

$A = l \cdot w$

$A = $ _____ • _____

$A = $ _____

The area is _____ in^2.

To find the area of a parallelogram, multiply the base b times the height h.
$$A = b \cdot h$$

Find the area of each parallelogram.

3.

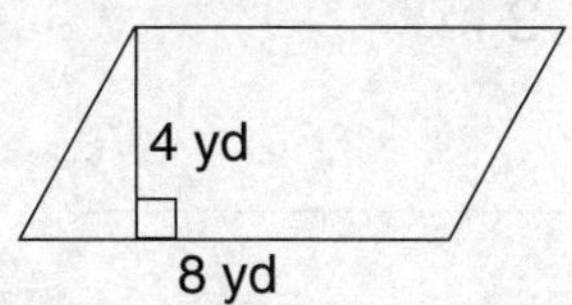
4 yd
8 yd

$A = b \cdot h$

$A = $ _____ • _____

$A = $ _____

The area is _____ yd^2.

4.

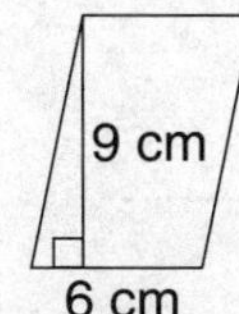
9 cm
6 cm

$A = b \cdot h$

$A = $ _____ • _____

$A = $ _____

The area is _____ cm^2.

Holt Mathematics

9-3

Challenge
Size It Up!

Gonzalez Builders purchased the four lots shown below. The company intends to build homes on the lots and needs to know the total area in square feet of each lot. They also need to know the perimeter in feet of each lot.

You can use proportions to calculate the areas and perimeters. For each figure, the scale is 1 inch: 60 feet.

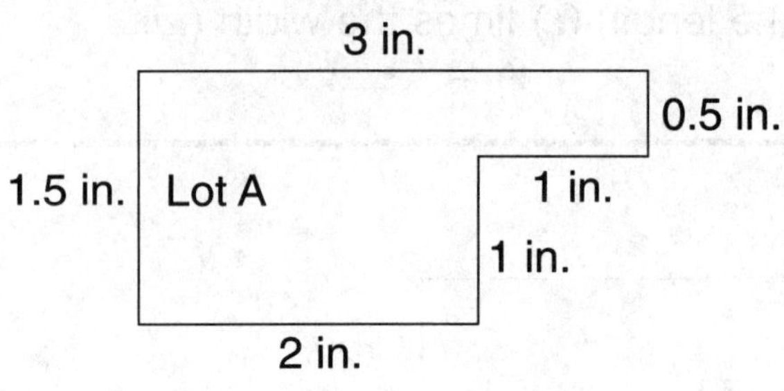

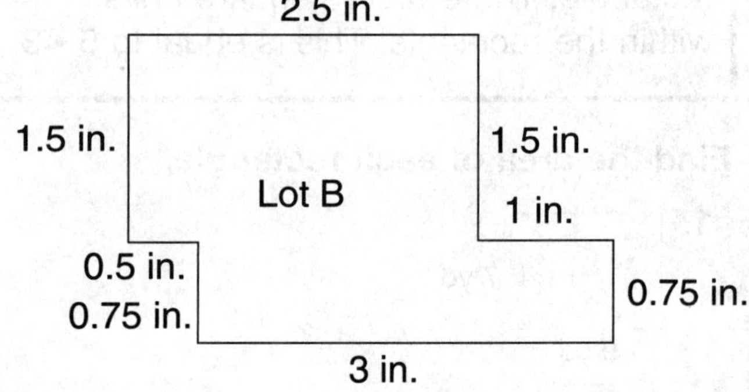

1. total area _________________

perimeter _________________

2. total area _________________

perimeter _________________

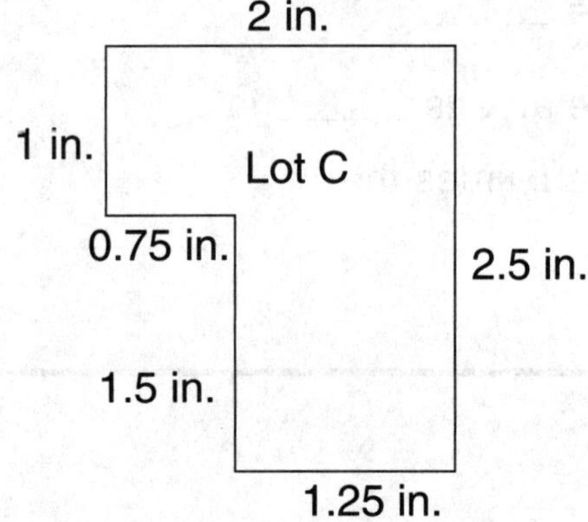

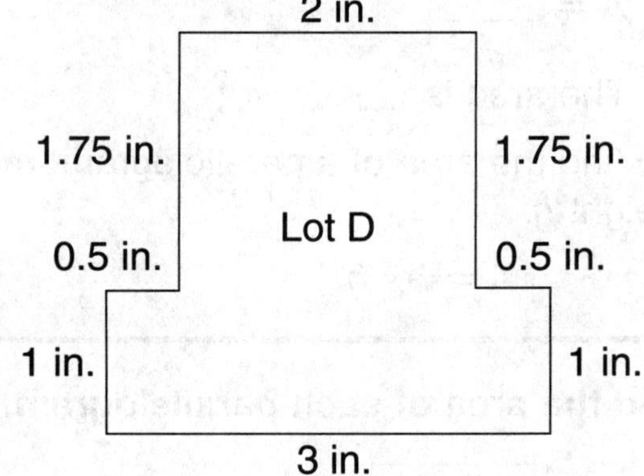

3. total area _________________

perimeter _________________

4. total area _________________

perimeter _________________

5. What is the combined area of all four lots? _____________________

6. What length of fencing would be needed to enclose all four lots during construction? _____________________

Holt Mathematics

Problem Solving

LESSON 9-3

Area of Parallelograms

Write the correct answer.

1. A dollar bill has an area of 15.86 square inches. If a dollar bill is 2.6 inches long, how wide is it?

2. On an official United States flag, the ratio of width to length is exactly 1 to 1.9. What is the area of a United States flag whose width is 2 feet?

3. A back yard is shaped like a parallelogram with a height of 32 feet and a base of 100 feet. One bag of grass seed covers 125 square feet. What is the least number of bags of seed needed to seed the lawn?

4. The art club is painting a mural on a school wall. The mural is in the shape of a parallelogram. If the base of the mural is 10.5 feet long and the mural covers 89.25 square feet, how high is the mural?

Choose the letter of the correct answer.

5. In baseball, the area of each base is 225 square inches. Each base is a square. What is the length of each side of a base on a baseball field?

   A 12 in. **C** 25 in.

   B 22.5 in. **D** 15 in.

6. The area of a parallelogram is 632.1 square centimeters. Its base is 24.5 centimeters. What is the height of the parallelogram?

   F 25.8 cm **H** 21.9 cm

   G 705.6 cm **J** 11.8 cm

7. The official rules for volleyball were developed in 1897. The rules state that the court or floor space must be 25 feet wide and 50 feet long. An official basketball court is 94 feet by 50 feet. How much larger is the area of a basketball court than the area of a volleyball court?

   A 69 ft^2 larger

   B 3,450 ft^2 larger

   C 1,250 ft^2 larger

   D 4,700 ft^2 larger

8. Two parallelograms each have an area of 288 square inches. One has a height of 12 inches, and the other has a height of 18 inches. What are the bases of each parallelogram?

   F 40 in. and 30 in.

   G 22 in. and 15 in.

   H 24 in. and 16 in.

   J 26 in. and 20 in.

Holt Mathematics

Reading Strategies
Compare and Contrast

Area measures the surface a figure covers.
Area is measured in square units.

10 cm

To find the area of this rectangle, count the number of square units.
There are eight square units.

Another way to find the area is to use this formula: $A = \ell w$.
Read: "Area equals length times width."

4 units times 2 units = 8 square units.
Area = 8 square units

Compare a parallelogram to a rectangle. The width of a
parallelogram is called the height. The length of a
parallelogram is called the base.

Moving one piece of the parallelogram
shows how the base and height compare
to the length and width of a rectangle.

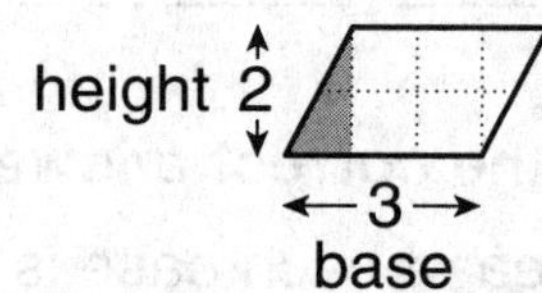

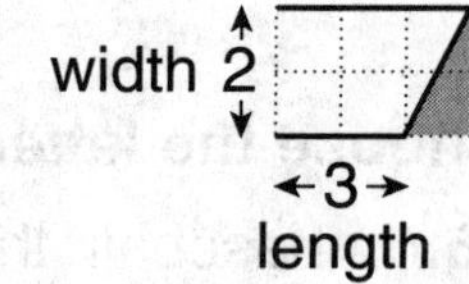

Area = base • height ($A = bh$)   Area = length • width ($A = \ell w$)
$A = 3 \cdot 2$ $A = 3 \cdot 2$
$A = 6$ square units $A = 6$ square units

1. The height of a parallelogram is the same as what part of a
   rectangle?

   ____

2. The base of a parallelogram is the same as what part of a
   rectangle?

   ____

3. What is the same about finding the area of a parallelogram and
   a rectangle?

   ____

   ____

4. What is different about finding the area of a parallelogram and a
   rectangle?

   ____

Holt Mathematics

Puzzles, Twisters & Teasers

LESSON 9-3

Reach the Heights!

Across

4. a four-sided figure with congruent and parallel opposite sides

5. shape

8. the width of the rectangle

Down

1. number of square units needed to cover the figure

2. parallelogram with four equal sides and angles

3. the length of the rectangle

6. _______ of measurement

7. a _________ board leads to "Checkmate"

Holt Mathematics

<table><tr><td>**LESSON 9-4**</td><td># Practice A
Area of Triangles and Trapezoids</td></tr></table>

Find the area of each triangle.

1. 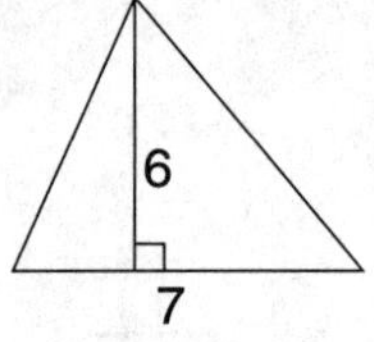

2. 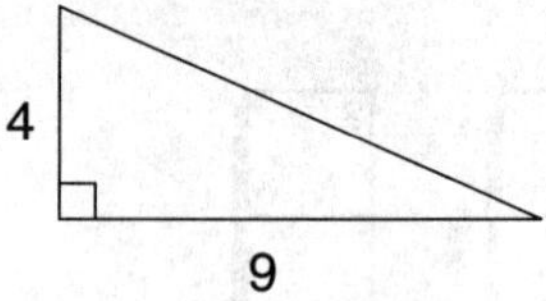

3. 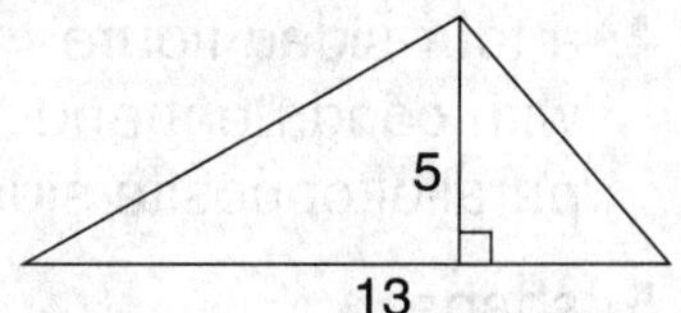

________________   ________________   ________________

Find the area of each trapezoid. Choose the letter for the best answer.

4. 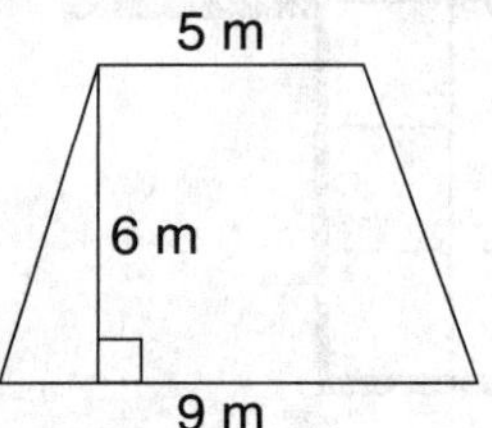

A 24 m^2	**C** 54 m^2
B 42 m^2	**D** 84 m^2

5. 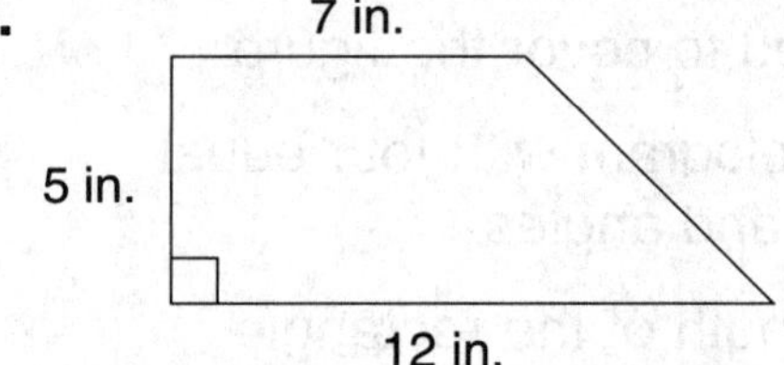

F 95 in^2	**H** 47.5 in^2
G 60 in^2	**J** 25 in^2

6. 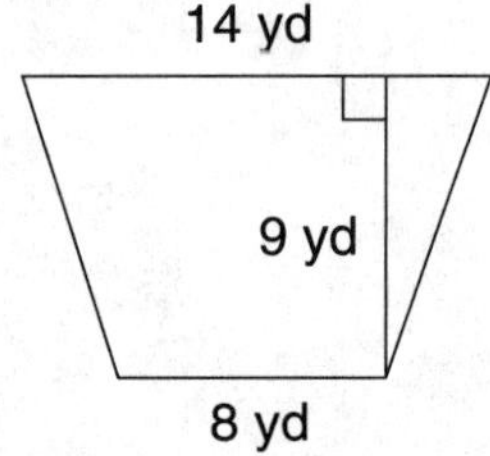

A 99 yd^2	**C** 198 yd^2
B 126 yd^2	**D** Not Here

7. 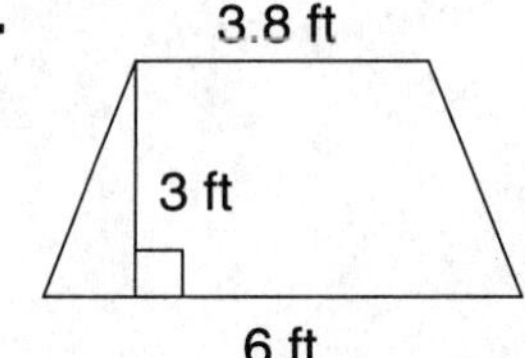

F 11.4 ft^2	**H** 18 ft^2
G 14.7 ft^2	**J** 29.4 ft^2

8. The state of Missouri is shaped somewhat like a trapezoid. What is the approximate area of Missouri?

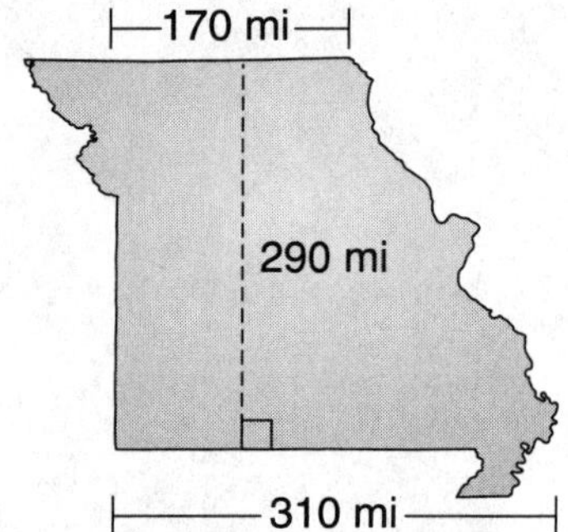

____

Holt Mathematics

Practice B
Area of Triangles and Trapezoids

Find the area of each triangle.

1. 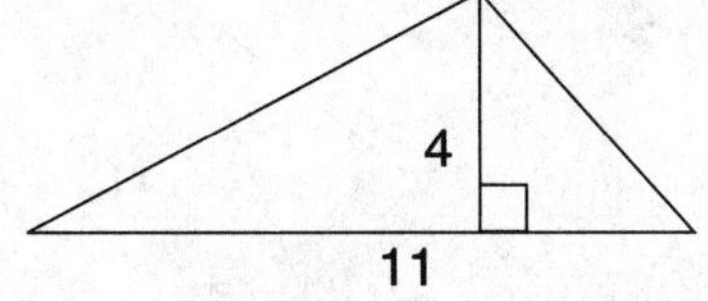

2. 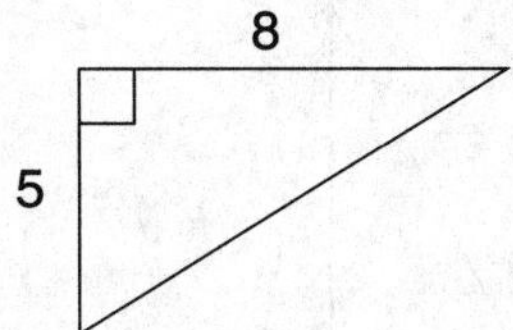

3. 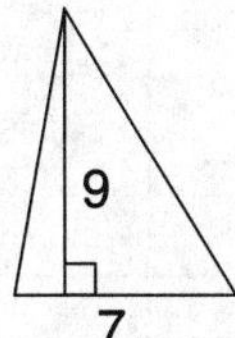

____

____

____

4. 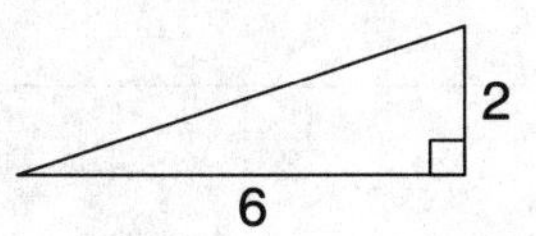

5. 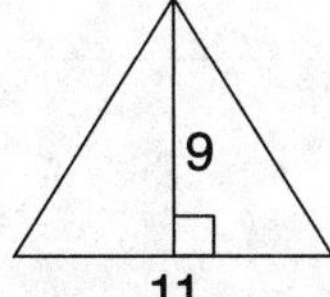

6. 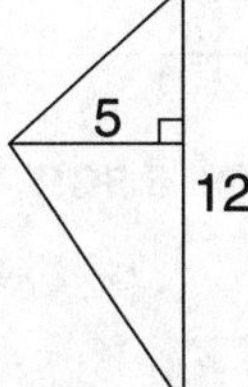

____

____

____

Find the area of each trapezoid.

7. 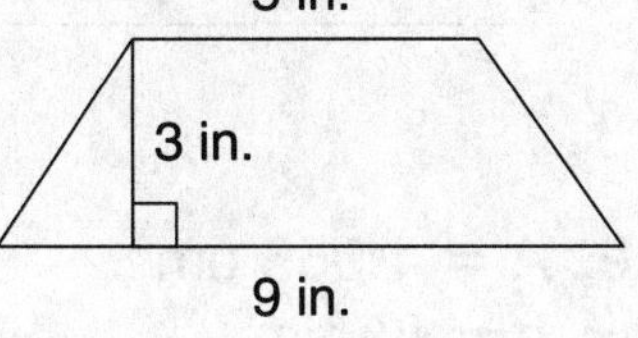

8. 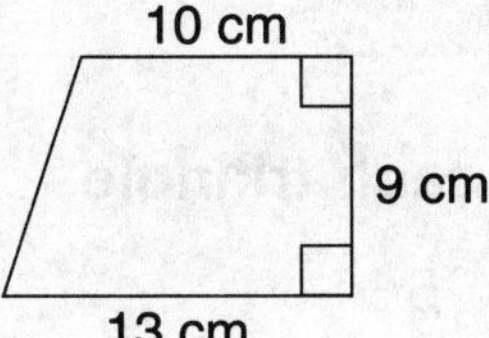

9. 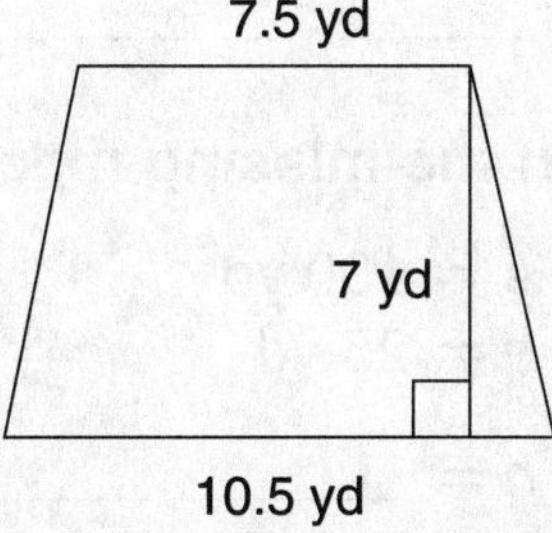

____

____

____

10. 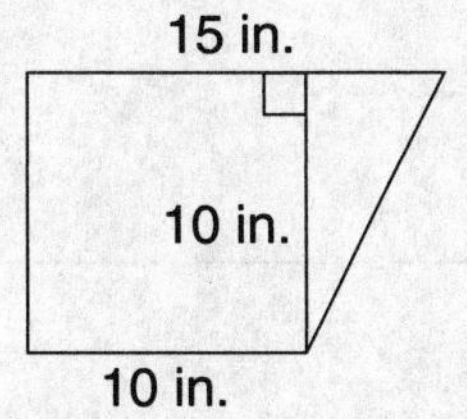

11. 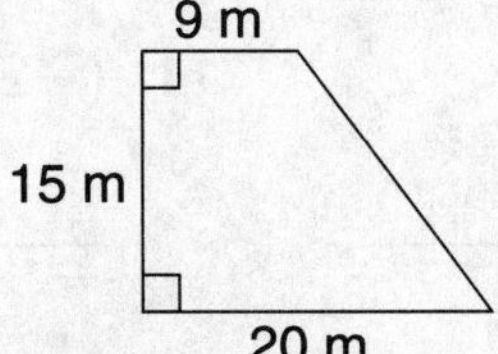

12. 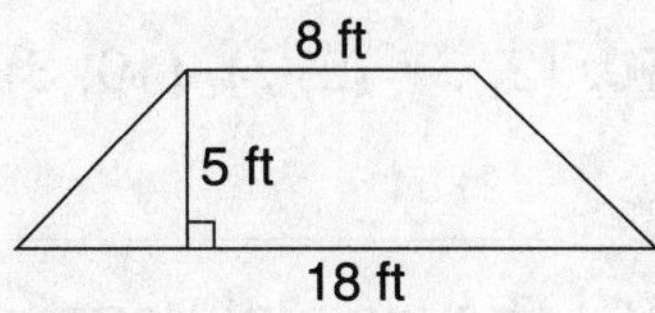

____

____

____

13. The state of Montana is shaped somewhat like a trapezoid. What is the approximate area of Montana?

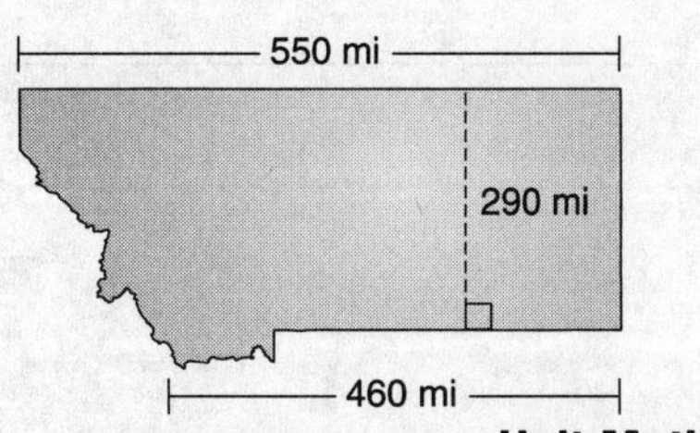

____

Holt Mathematics

LESSON 9-4 Practice C
Area of Triangles and Trapezoids

Find the area of each triangle.

1. 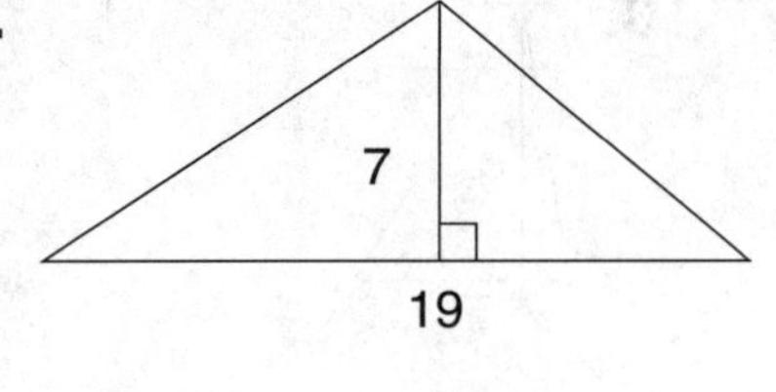

2. 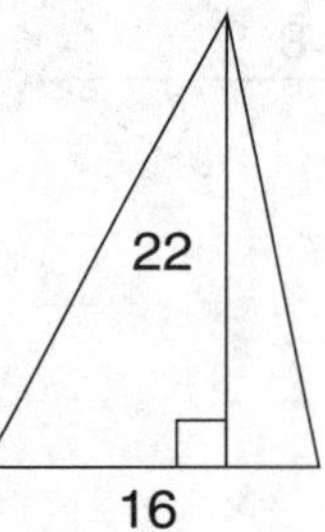

3. 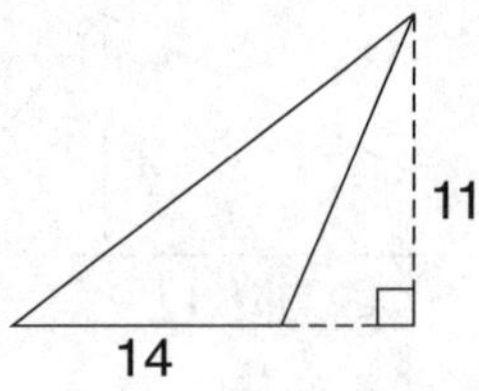

_______________ _______________ _______________

Find the area of each trapezoid.

4. 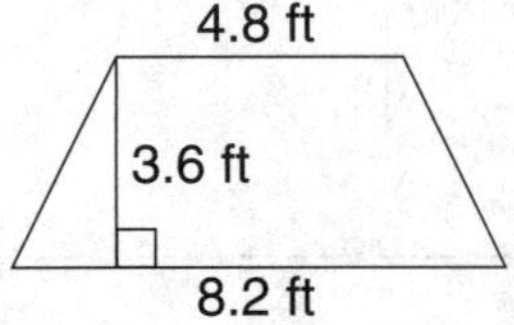

5. 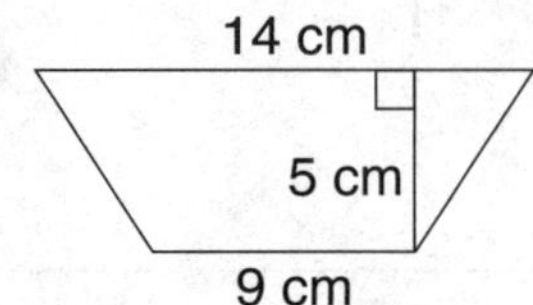

6. 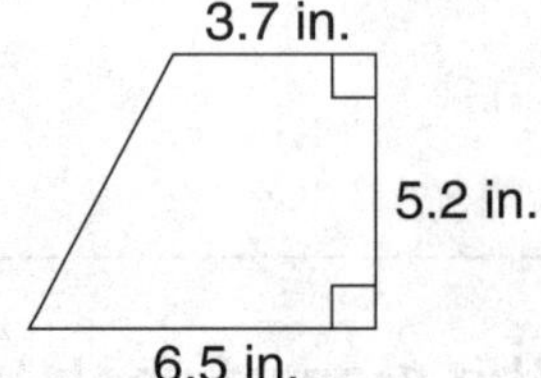

_______________ _______________ _______________

Find the missing measurement of each triangle.

7. $A = 100 \text{ yd}^2$
$b = 25 \text{ yd}$

$h =$ _______________

8. $b = 5 \text{ in.}$
$h = 0.8 \text{ in.}$

$A =$ _______________

9. $A = 1{,}955 \text{ cm}^2$
$h = 85 \text{ cm}$

$b =$ _______________

Graph the polygon with the given vertices. Then find the area of the polygon.

10. (2, 3), (5, 7), (10, 3), (9, 7)

11. (−2, 6), (−7, 1), (−7, 6)

_______________ _______________

12. The state of Vermont is shaped somewhat like a trapezoid. What is the approximate area of Vermont?

____

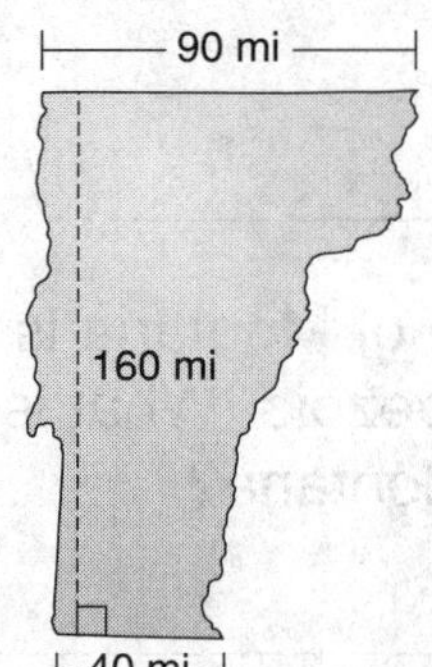

Holt Mathematics

LESSON 9-4 · **Reteach**
Area of Triangles and Trapezoids

The diagram shows how you can cut a parallelogram into two congruent triangles.

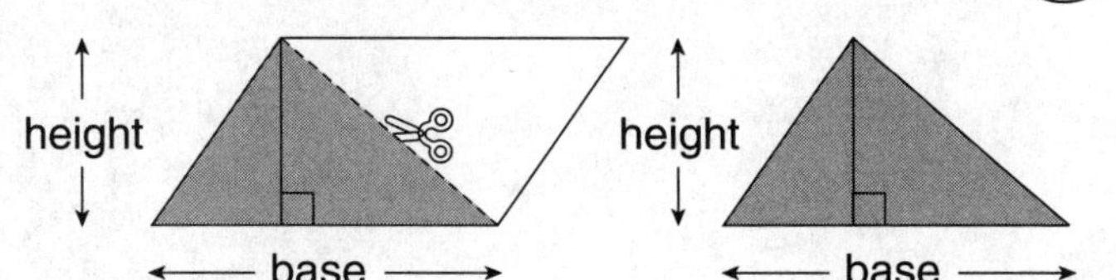

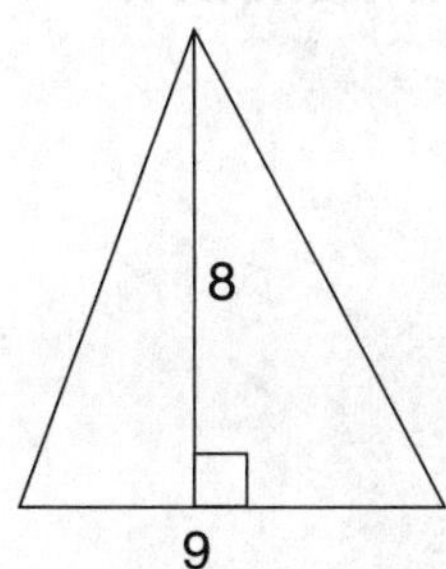

The area of the triangle is $\frac{1}{2}$ the area of the parallelogram.

The formula for the area of a triangle is $A = \frac{1}{2} \cdot b \cdot h$.

Find the area of each triangle.

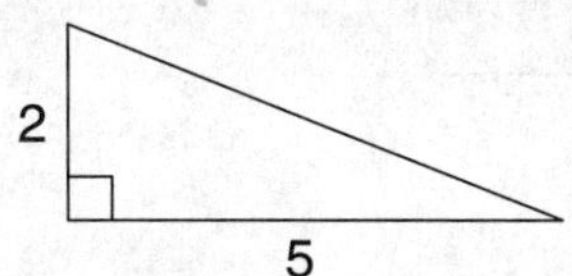

1. $A = \frac{1}{2} \cdot b \cdot h$

$A = \frac{1}{2} \cdot \underline{\quad} \cdot \underline{\quad}$

$A = \frac{1}{2} \cdot \underline{\quad}$

$A = \underline{\quad}$

The area of the triangle is ____ units^2.

2. $A = \frac{1}{2} \cdot b \cdot h$

$A = \frac{1}{2} \cdot \underline{\quad} \cdot \underline{\quad}$

$A = \frac{1}{2} \cdot \underline{\quad}$

$A = \underline{\quad}$

The area of the triangle is ____ units^2.

3.

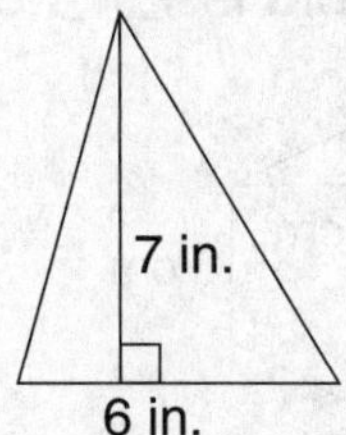

4.

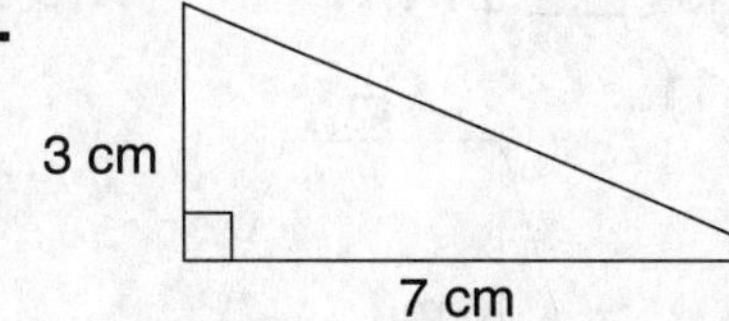

5.

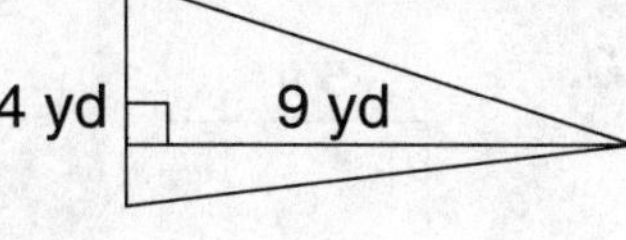

____________________   ____________________

6. What is the area of a triangle with base 16 m and height 10 m? ____________

7. What is the area of a triangle with base 25 mm and height 50 mm? ____________

Holt Mathematics

LESSON 9-4 **Reteach**
Area of Triangles and Trapezoids (continued)

In a trapezoid, the parallel sides are called the *bases*. One base is always longer than the other. The bases are labeled base 1 and base 2.

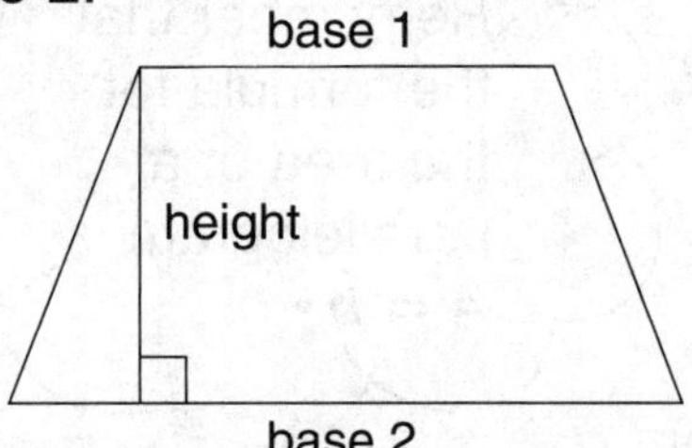

Area of trapezoid $= \frac{1}{2}h(b_1 + b_2)$

Find the area of each trapezoid.

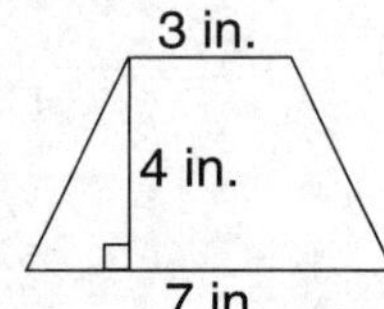

8. $A = \frac{1}{2}h(b_1 + b_2)$

$A = \frac{1}{2} \cdot \underline{\quad}(\underline{\quad} + \underline{\quad})$

$A = \frac{1}{2} \cdot \underline{\quad}(\underline{\quad})$

$A = \frac{1}{2} \cdot \underline{\quad}$

$A = \underline{\quad}$

The area of the trapezoid is ____ in^2.

9. $A = \frac{1}{2}h(b_1 + b_2)$

$A = \frac{1}{2} \cdot \underline{\quad}(\underline{\quad} + \underline{\quad})$

$A = \frac{1}{2} \cdot \underline{\quad}(\underline{\quad})$

$A = \frac{1}{2} \cdot \underline{\quad}$

$A = \underline{\quad}$

The area of the trapezoid is ____ cm^2.

10.

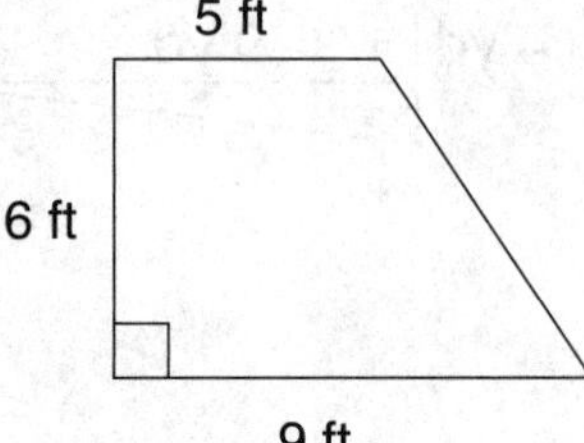

11.

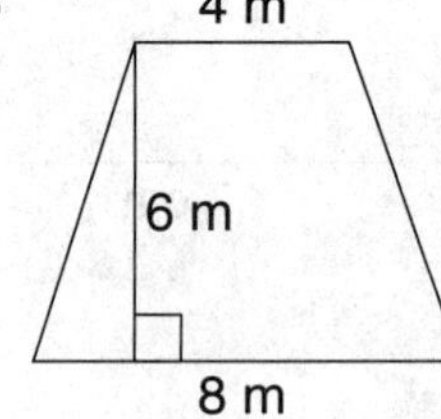

12.

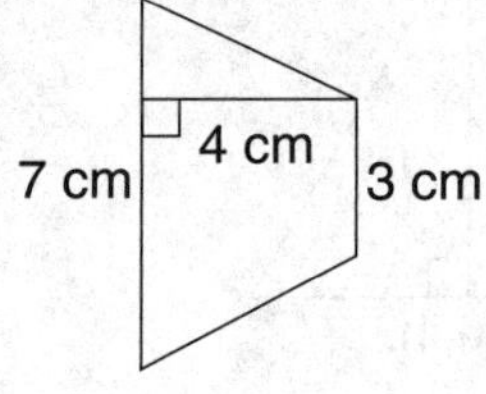

____________ ____________ ____________

13. What is the area of a trapezoid with bases 25 yd and 75 yd and height 10 yd?

____

Holt Mathematics

LESSON 9-4 Challenge
Break It Up

One way to find the area of this figure is to divide it into a rectangle
and a triangle. Find the area of each, then add the areas together.

Area of rectangle = $7 \cdot 10 = 70$ m^2

Area of triangle = $\frac{1}{2}(5 \cdot 10) = 25$ m^2

Total area of figure = Area of rectangle + Area of triangle

$$= 70 \text{ m}^2 + 25 \text{ m}^2 = 95 \text{ m}^2$$

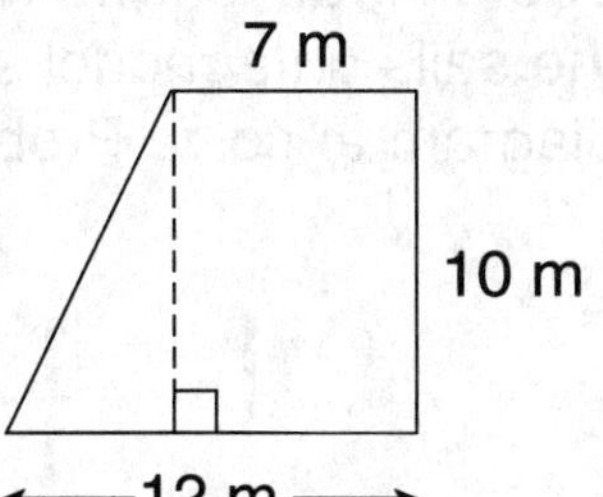

**Divide each figure into parts. Then find the area of each part.
Add the areas of the parts to find the area of the whole figure.**

1.

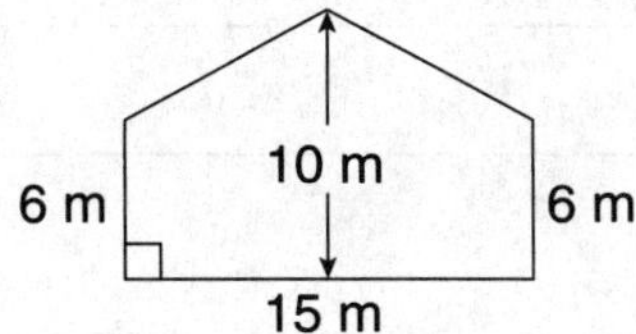

2.

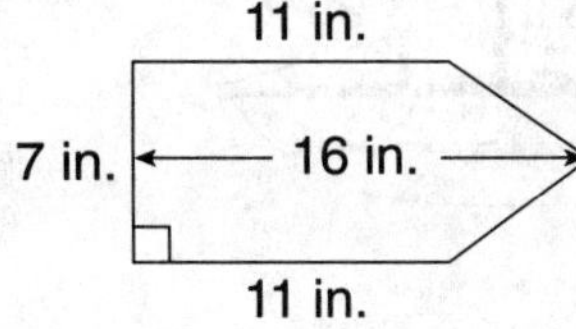

____

3.

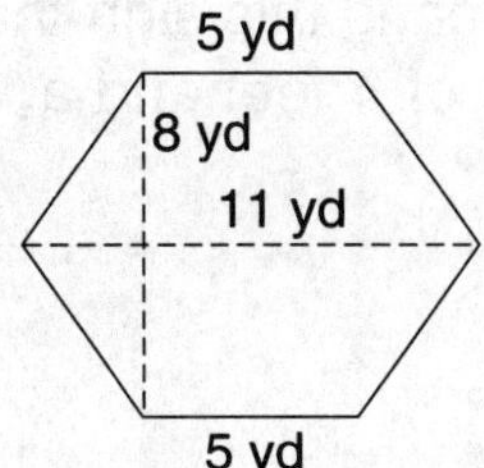

4.

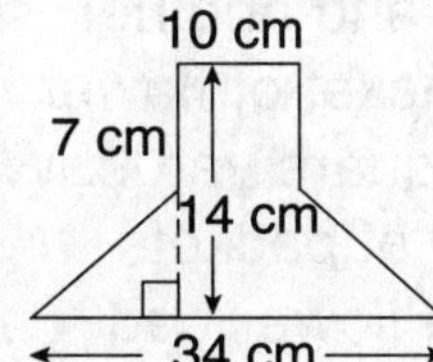

____

**For Exercises 5 and 6, you need to *subtract* the area of a part
to find the area of each whole figure.**

5.

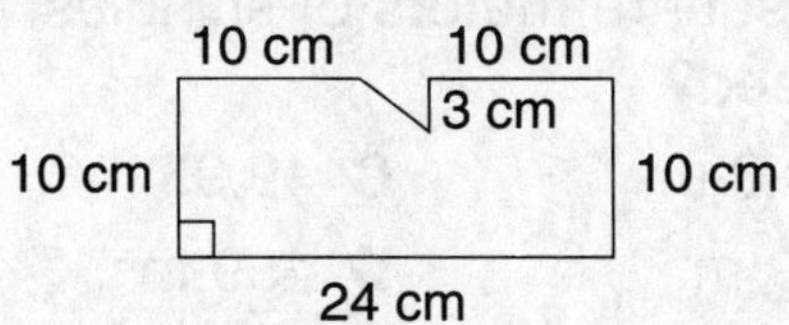

6.

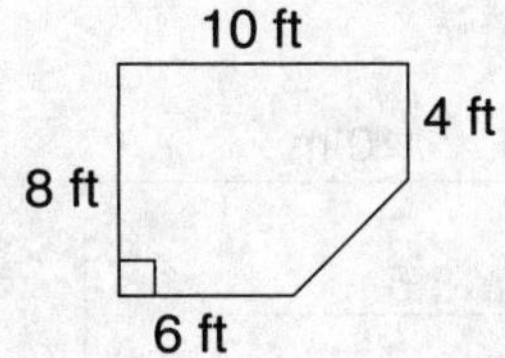

____

Holt Mathematics

Name _______________________________ Date __________ Class __________

LESSON 9-4 Problem Solving
Area of Triangles and Trapezoids

Write the correct answer.

The diagram shows the dimensions of the sails on a model sailboat. Use the diagram to solve Problems 1–2.

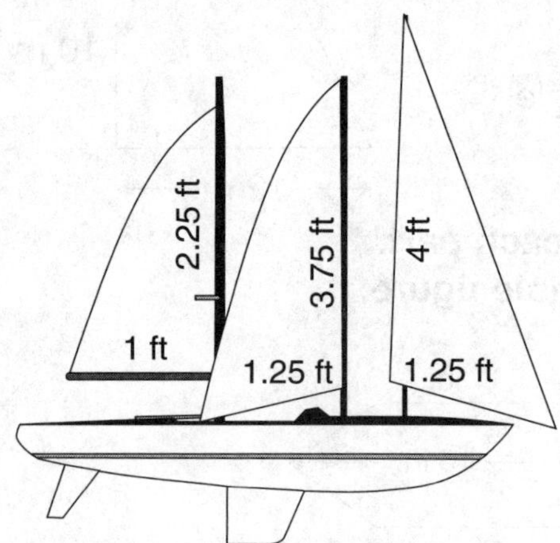

1. About how much material to the nearest square foot will be needed to make the sails?

2. If the dimensions for each sail were doubled, how would that change the total amount of material needed to make the sails?

3. A flower bed is shaped like a trapezoid with a height of 3.5 yards, one 2.8-yard base, and another 4.6-yard base. A packet of flower seeds covers 5.6 square yards. What is the least number of packets needed to plant the flower bed?

4. A triangular road sign has a height of 8 feet and a base of 16.5 feet. How much larger in area is this sign than one with a height of 4 feet and a base of 8.25 feet?

Choose the letter of the correct answer.

This diagram shows the top view of the roof of a house.

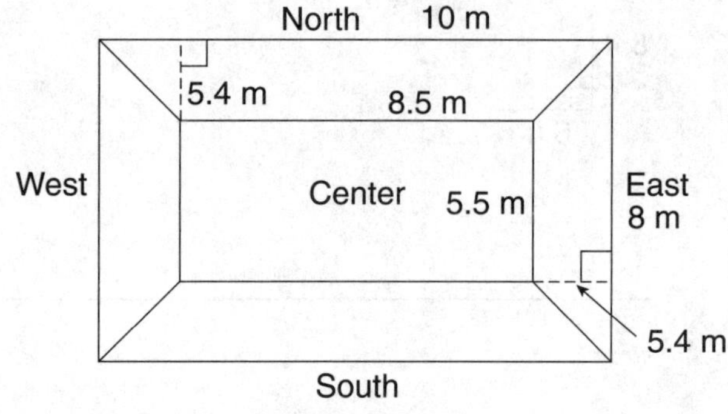

5. If you need to reshingle the north and south sections of the roof, how many square meters of shingles will you need?
   A 199.8 m^2 C 49.95 m^2
   B 99.9 m^2 D 459 m^2

6. If you need to reshingle the west section of the roof, how many square meters of shingles will you need?
   F 13.5 m^2 H 36.45 m^2
   G 18.9 m^2 J 72.9 m^2

Holt Mathematics

LESSON 9-4 — **Reading Strategies**
Use Graphic Aid

Knowing how to find the area of a square or parallelogram can help you find the area of a triangle.

When you draw a diagonal line through a square, you make two equal triangles.

Answer each question.

1. How many square units are in the square? ____________

2. To find out how many squares are in one of the triangles, count the number of half squares first. How many did you find? ____________

3. How many whole squares can you count? ____________

4. How many half and whole squares together? ____________

You can also draw a diagonal line through a parallelogram and divide it into two equal triangles.

Use the figure to answer each.

5. How many square units are in the parallelogram? ____________

6. Count the full squares and half squares in one triangle and write the total. ____________

7. Compare the area of a parallelogram or square with the area of a triangle.

____

Holt Mathematics

LESSON 9-4	**Puzzles, Twisters & Teasers**
	All Washed Up!

Find the area of each figure below. Match the letters to solve the riddle.

1. ___________ S

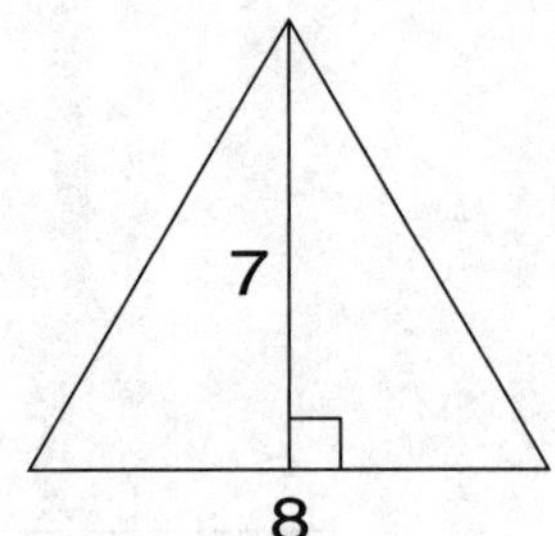

2. ___________ M

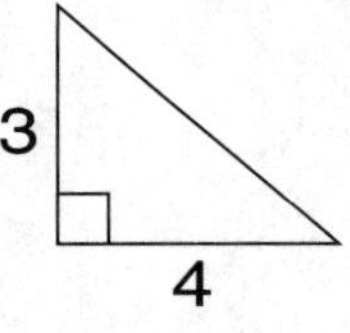

3. ___________ E

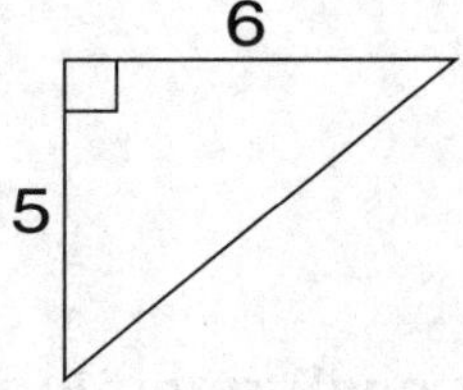

4. ___________ I

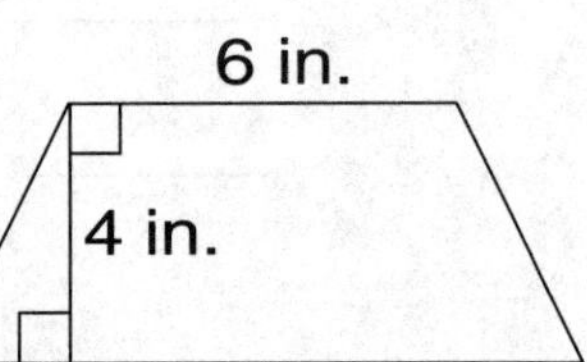

5. ___________ V

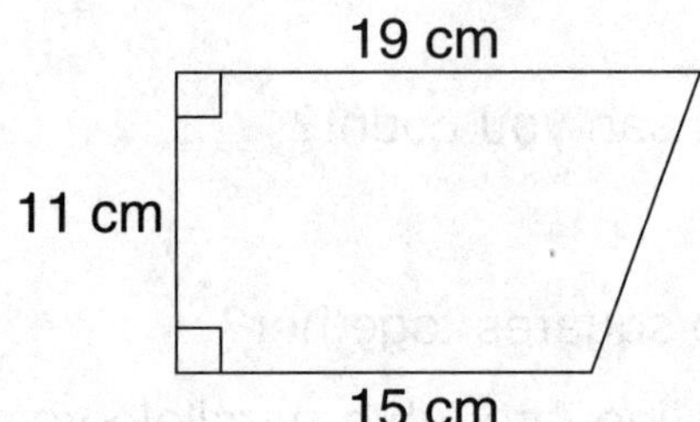

6. ___________ C

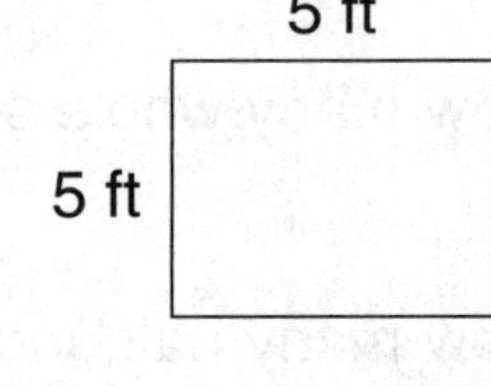

7. ___________ A

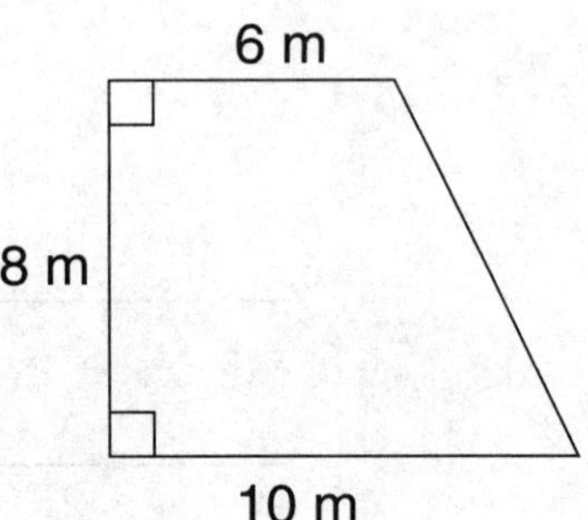

8. ___________ R

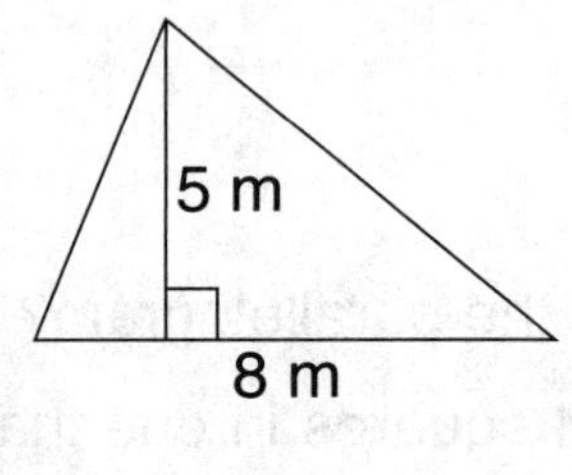

9. ___________ W

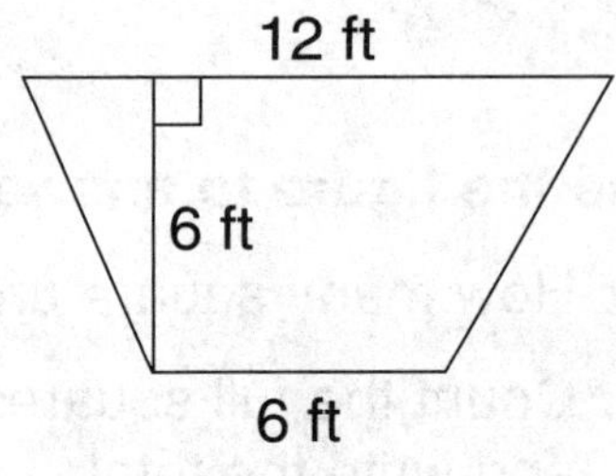

What washes up on small beaches?

__ __ __ __ 0 - __ __ __ __ __
6 32 25 20   54 64 187 15 28

Holt Mathematics

LESSON 9-5 Practice A
Area of Circles

Find the area of each circle to the nearest tenth. Use 3.14 for π.
Cross out each number in the box that matches an area.

19.6	254.3	50.2	78.5	22.0	260.0
153.9	28.3	3.1	379.9	171.9	72.3 176.6

1.

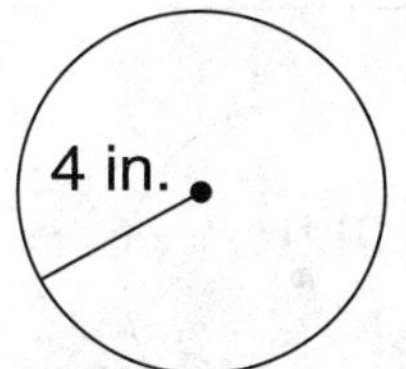

____

2.

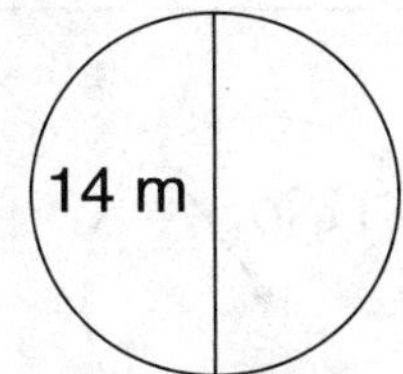

____

3.

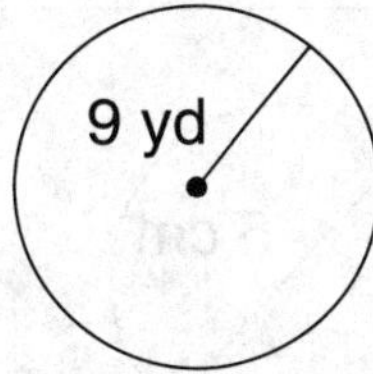

____

4.

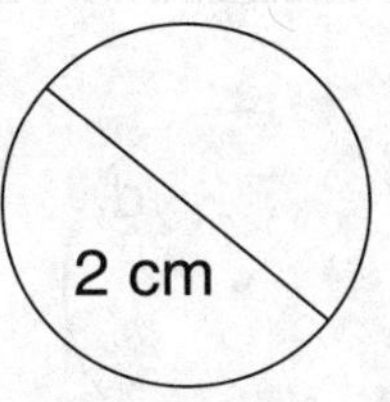

____

5.

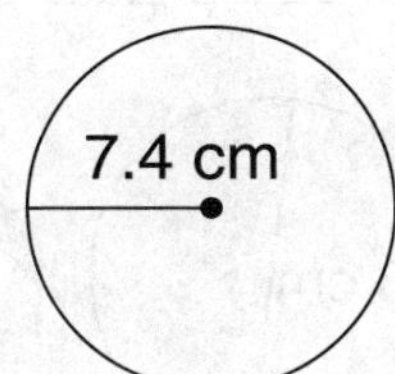

____

6.

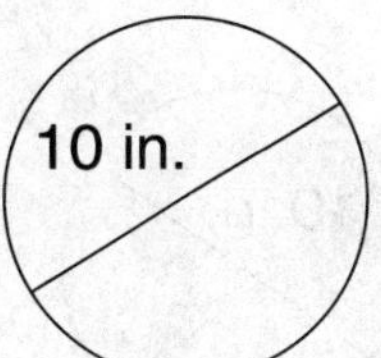

____

7.

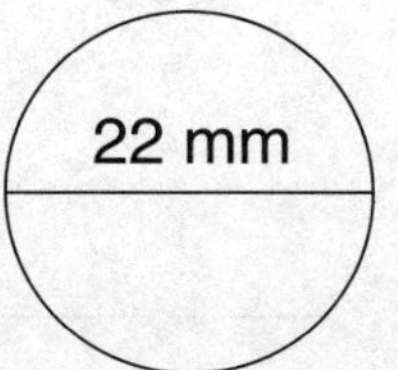

____

8.

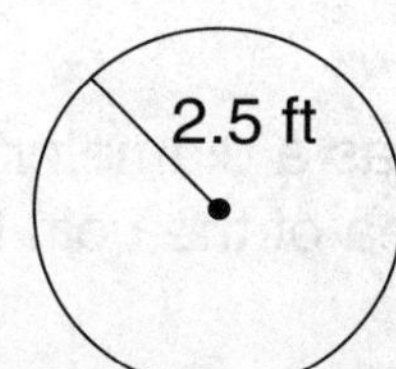

____

9.

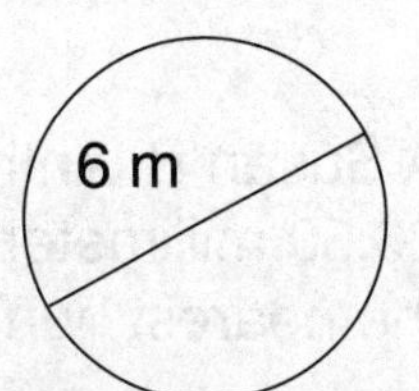

____

10. A round table has a diameter of 28 inches.
What is the area of the tabletop? Use $\frac{22}{7}$ for π. ______________________

11. Find the area of the shaded region of the
circle. Use 3.14 for π. Round your answer
to the nearest tenth.

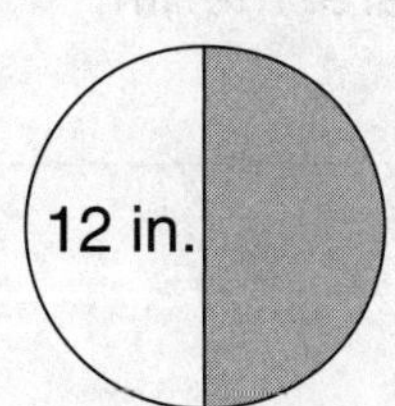

____

Holt Mathematics

Practice B
LESSON 9-5
Area of Circles

Find the area of each circle to the nearest tenth. Use 3.14 for π.

1. 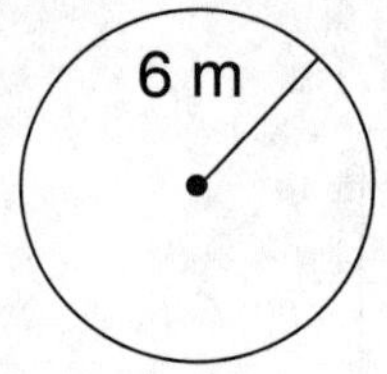

____

2. 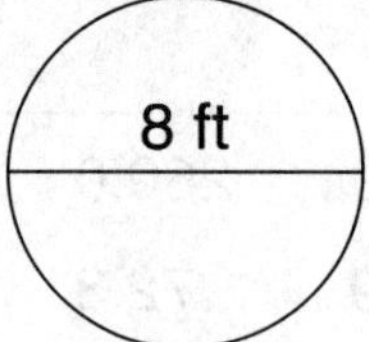

____

3. 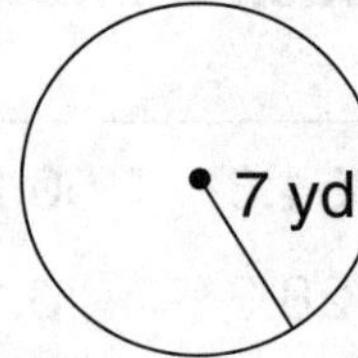

____

4. 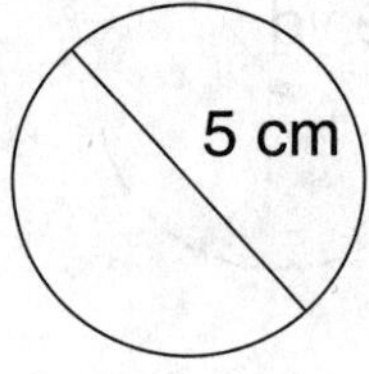

____

5. 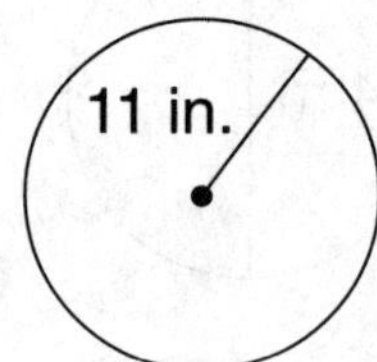

____

6. 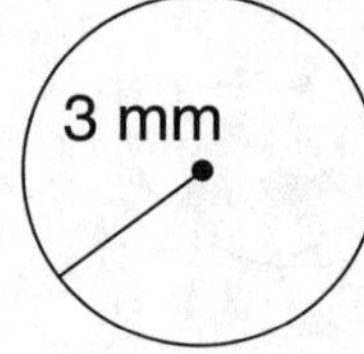

____

7. 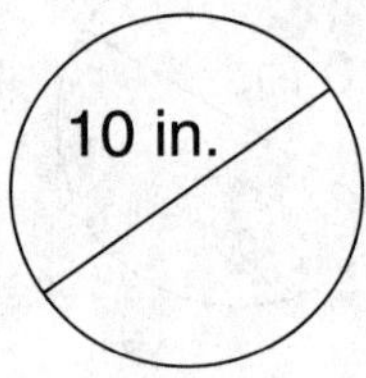

____

8. 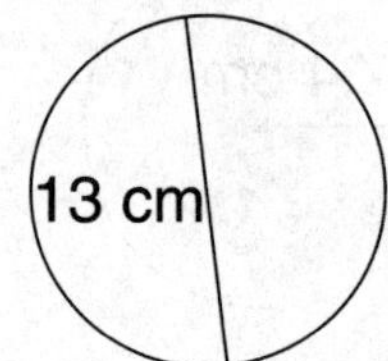

____

9. 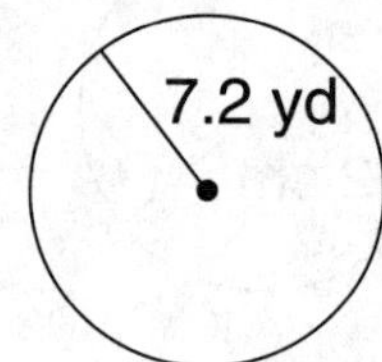

____

10. A Susan B. Anthony dollar coin has a diameter of 26.50 millimeters. What is the area of the coin to the nearest hundredth?

____

11. A tablecloth for a round table has a radius of 21 inches. What is the area of the tablecloth? Use $\frac{22}{7}$ for π.

____

12. Use a centimeter ruler to measure the radius of the circle. Then find the area of the shaded region of the circle. Use 3.14 for π. Round your answer to the nearest tenth.

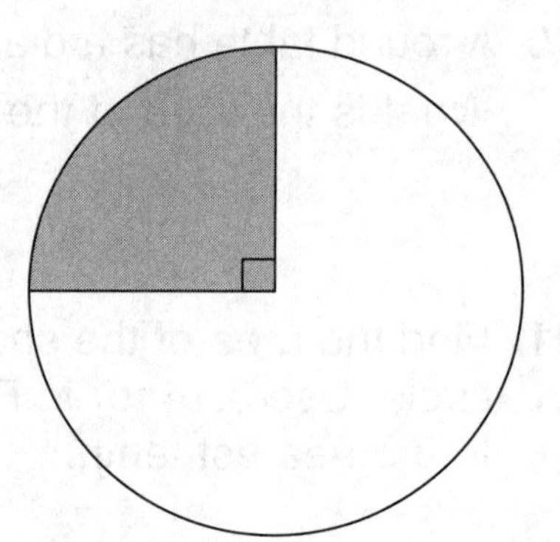

____

Holt Mathematics

Practice C
Area of Circles

Find the area of each circle to the nearest tenth. Use 3.14 for π.

1. 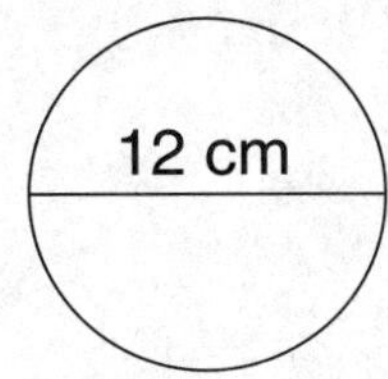

____

2. 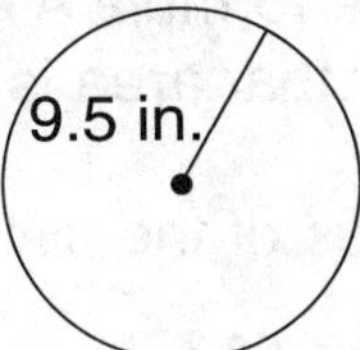

____

3. 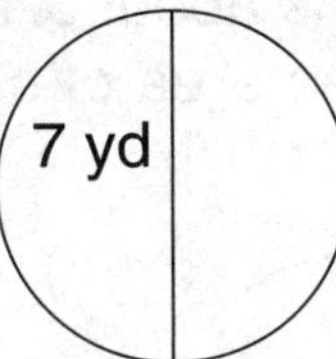

____

4. 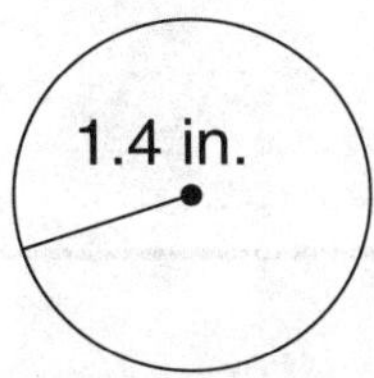

____

5. 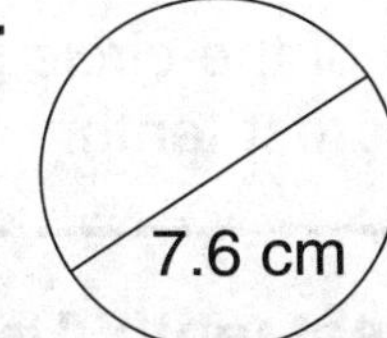

____

6. 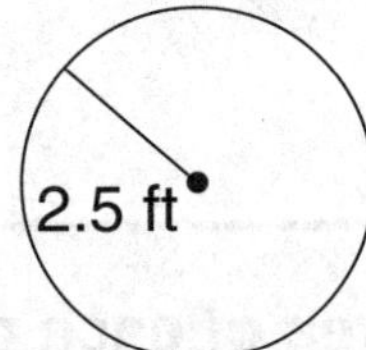

____

Given the radius or diameter, find the circumference and area of each circle to the nearest tenth. Use 3.14 for π.

7. $r = 4$ in.

8. $d = 14$ cm

9. $d = 26$ ft

____

Given the area, find the radius of each circle. Use 3.14 for π.

10. $A = 200.96$ m^2

11. $A = 12.56$ in^2

12. $A = 314$ yd^2

____

13. A playground area is circular with a diameter of 32 feet. What is the area of the playground? Round your answer to the nearest tenth.

____

14. Use a centimeter ruler to measure the radius of the circle. Then find the area of the shaded region of the circle. Use 3.14 for π. Round your answer to the nearest tenth.

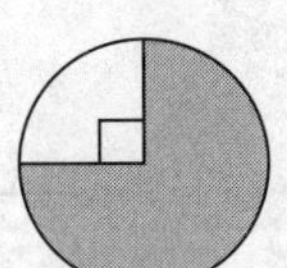

____

Holt Mathematics

LESSON 9-5

Reteach
Area of Circles

The formula $A = \pi r^2$ is used to find the area of a circle. Since the value of π is about 3.14, you can use the formula $A \approx 3.14 \cdot r \cdot r$ to estimate the area of a circle. Remember that area is expressed in square units.

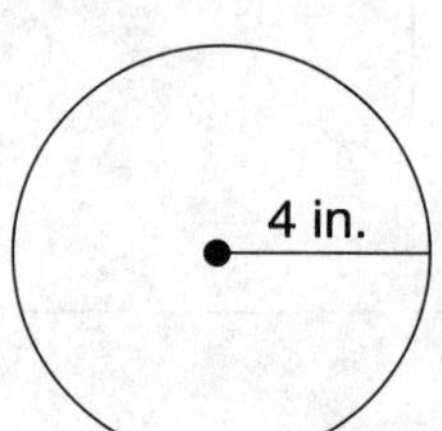

The radius of the circle is 4 in.

$A \approx 3.14 \cdot r \cdot r$

$A \approx 3.14 \cdot 4 \cdot 4$

$A \approx 50.24$

The area of the circle is 50.2 in^2 to the nearest tenth.

Find the area of each circle to the nearest tenth. Use 3.14 for π.

1.

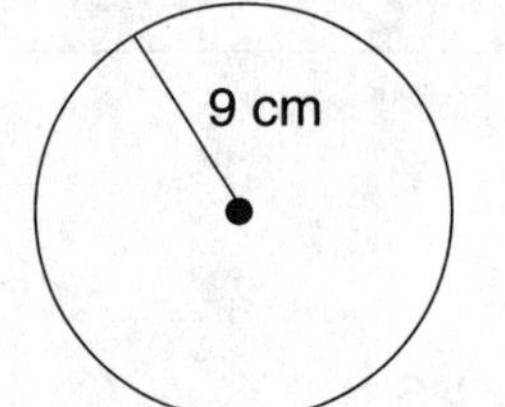

The radius is _______ cm.

$A = \pi r^2$

$A \approx 3.14 \cdot$ _______ $\cdot$ _______

$A \approx$ _____________

The area is _____________ cm^2 to the nearest tenth.

2.

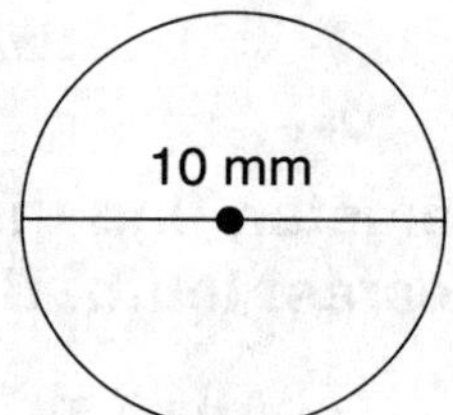

The diameter is 10 mm.

The radius is _______ mm.

$A = \pi r^2$

$A \approx 3.14 \cdot$ _______ $\cdot$ _______

$A \approx$ _____________

The area is _____________ m^2 to the nearest tenth.

3.

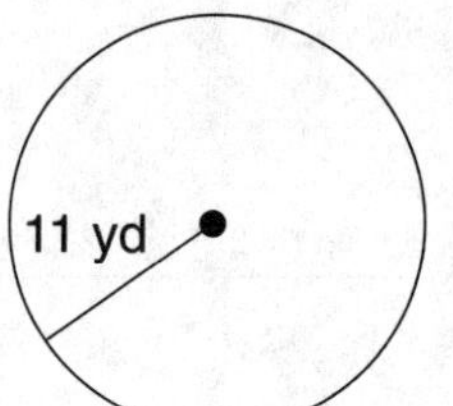

____

4.

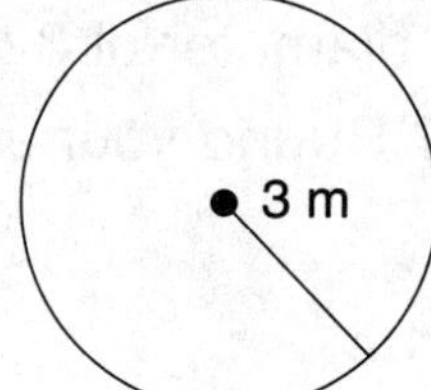

____

5.

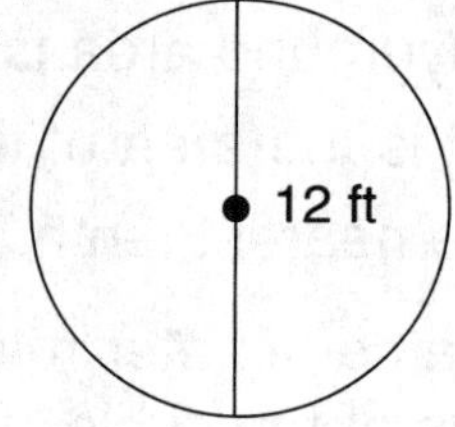

____

6. What is the area of a circle with radius 13 yd?
Round your answer to the nearest tenth. ___________________

Holt Mathematics

<table><tr><td>**LESSON**
9-5</td><td>

Challenge
Part of the Picture
</td></tr></table>

You can add or subtract to find the area of a shaded region.

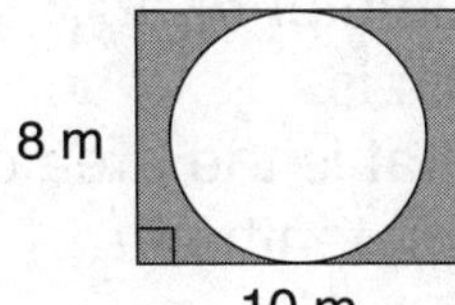

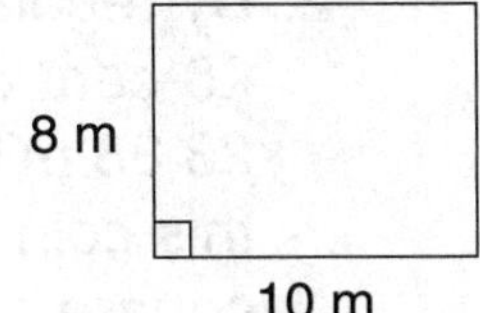

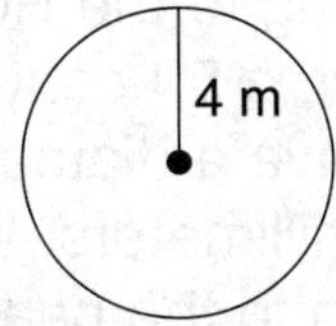

Find the area of the rectangle.
$$A = \ell w = 10 \cdot 8 = 80 \text{ m}^2$$

Find the area of the circle.
$$A = \pi r^2 = \pi \cdot 4^2 \approx 50.24 \text{ m}^2$$

The shaded area $\approx 80 \text{ m}^2 - 50.24 \text{ m}^2 \approx 29.76 \text{ m}^2 \approx 29.8 \text{ m}^2$.

Add or subtract to find the area of the shaded region.
Round your answer to the nearest tenth.

1.

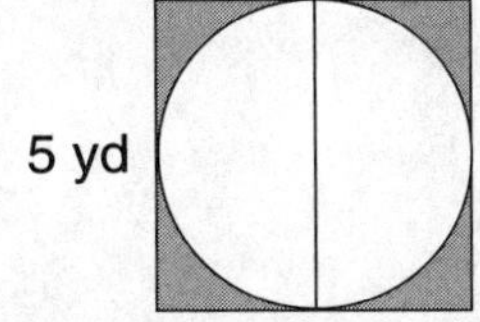

2.

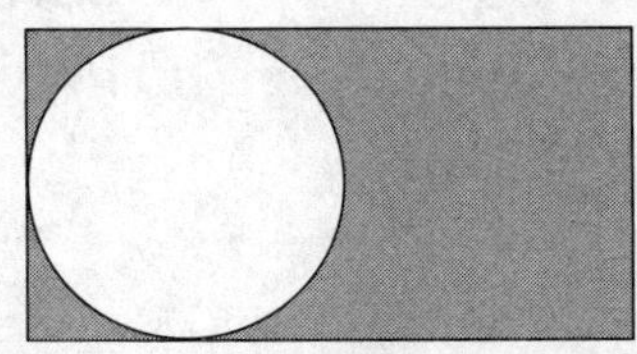

3.

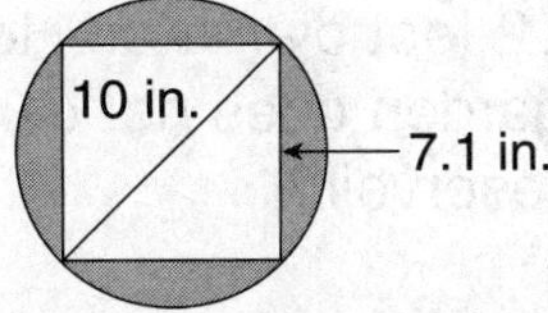

_______________ _______________ _______________

4.

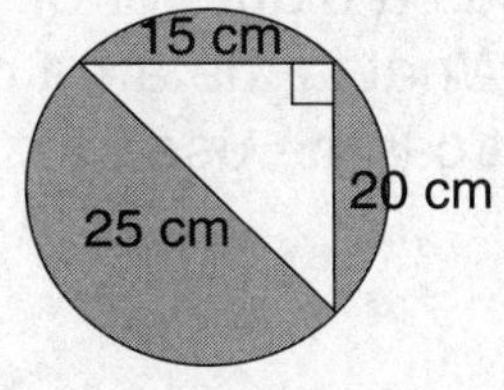

5.

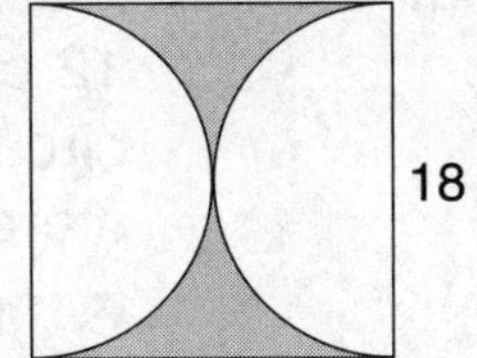

6.

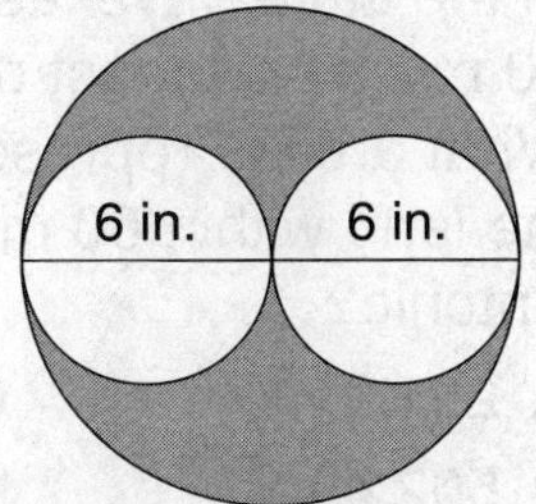

_______________ _______________ _______________

7. 8 m

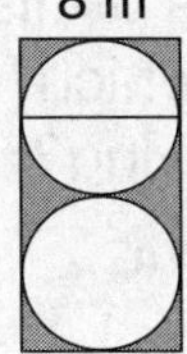

8.

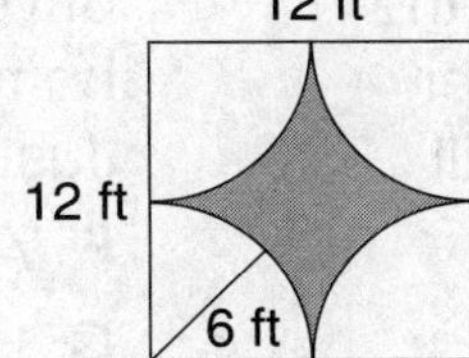

9.

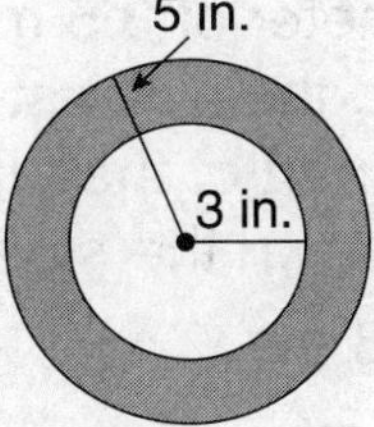

_______________ _______________ _______________

Holt Mathematics

Problem Solving
Area of Circles

Write the correct answer.

1. According to the Royal Canadian Mint Act, a 50-cent Canadian coin must have a diameter of 27.13 millimeters. What is the area of this coin to the nearest tenth of a square millimeter?

2. By regulation, the diameter of a 25-cent Canadian coin is 23.88 millimeters. What is the area of this coin to the nearest tenth of a square millimeter?

3. There is a water reservoir beneath a circular garden to supply a fountain in the garden. The reservoir has a 26-inch diameter. The garden has a 12-foot diameter. How much of the garden does not contain the water reservoir?

4. A frying pan has a diameter of 11 inches. What is the area to the nearest square inch of the smallest cover that will fit on top of the frying pan?

Choose the letter of the correct answer.

5. In the state of Texas, Austin is about 80 miles northeast of San Antonio. What area is represented by all of the land within 80 miles of San Antonio?

   A 251.2 mi^2 **C** 5,024 mi^2

   B 502.4 mi^2 **D** 20,096 mi^2

6. A standard CD has a diameter of 12 centimeters. What is the area of a circular case that can be used to store a CD?

   F 114 cm^2 **H** 105 cm^2

   G 112 cm^2 **J** 92 cm^2

7. A round dining table has a diameter of 2.5 meters. A round tablecloth has a diameter of 3.5 meters. What is the area to the nearest tenth of a meter of the part of the tablecloth that will hang down the side of the table?

   A 18.8 m^2 **C** 4.7 m^2

   B 6.3 m^2 **D** 1.0 m^2

8. Justin just got his driver's license. His parents are giving him permission to drive within a 25-mile radius of his home. What is the area to which Justin is restricted when driving?

   F 7,850 mi^2 **H** 157 mi^2

   G 1,962.5 mi^2 **J** 314 mi^2

Holt Mathematics

LESSON 9-5 — Reading Strategies
Use Graphic Aid

You can use what you know about the area of parallelograms to understand the area of circles.

The formula for the area of a parallelogram is:

Area = bh ⟶

height

base

You can cut a circle into wedges and place the wedges next to each other to approximate a parallelogram.

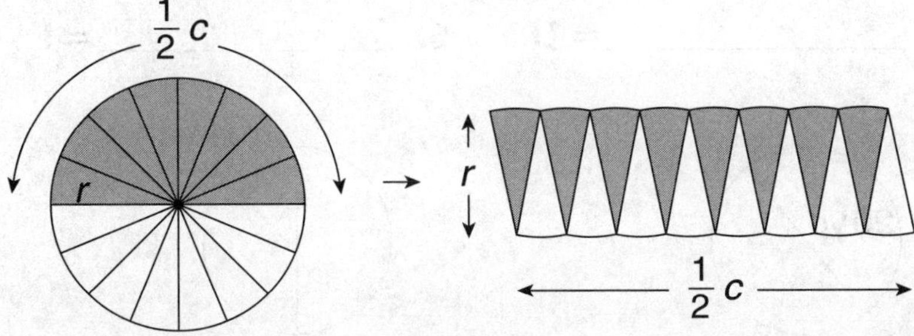

Compare the parallelogram to the circle to answer each question.

1. What figure do the wedges of a circle approximately form?

____

2. What is the formula for finding the area of a parallelogram?

____

3. A parallelogram has a measurement called the *base.* When a circle is reshaped into a parallelogram, what measurement is the same as the base?

____

4. The other measurement of a parallelogram is the height. When a circle is reshaped into a parallelogram, what measurement is the same as the height?

____

5. Use the information from Exercises 3 and 4 and the rearranged parallelogram to write the formula for the area of a circle.

____

Holt Mathematics

Puzzles, Twisters & Teasers
Area of Circles

**Find the area of each circle and write it on the line.
Use 3.14 for π. Then solve the riddle.**

1. _____________ = E   2. _____________ = T   3. _____________ = S

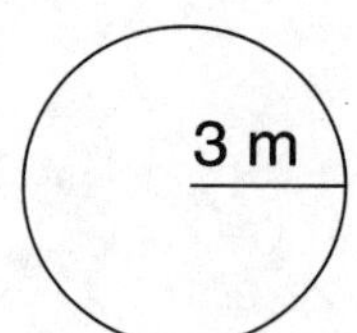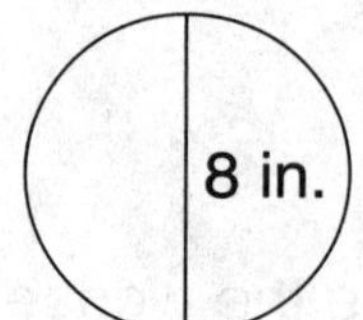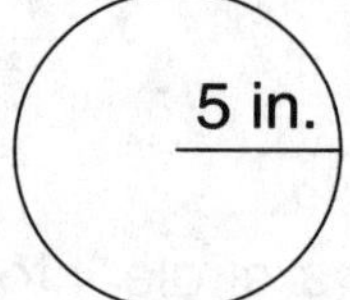

4. _____________ = L   5. _____________ = U   6. _____________ = I

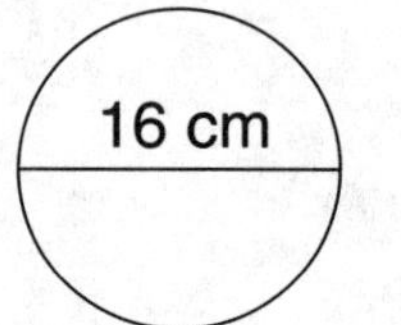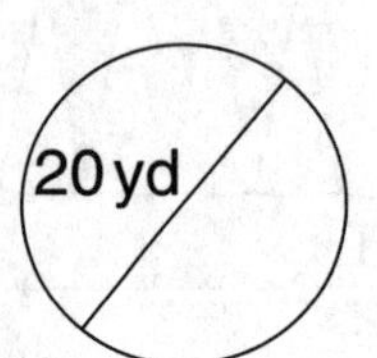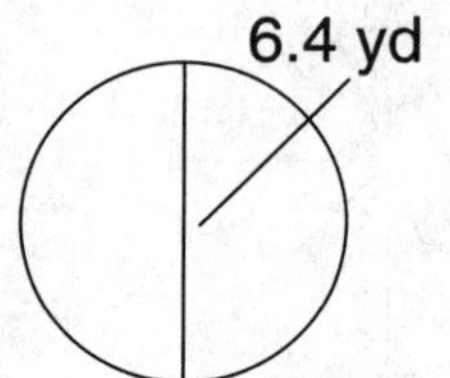

7. _____________ = O   8. _____________ = G   9. _____________ = H

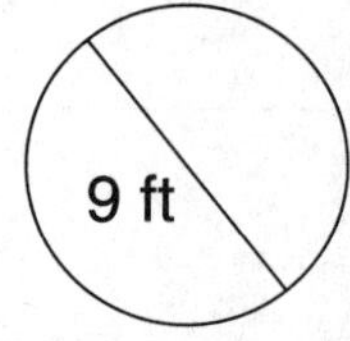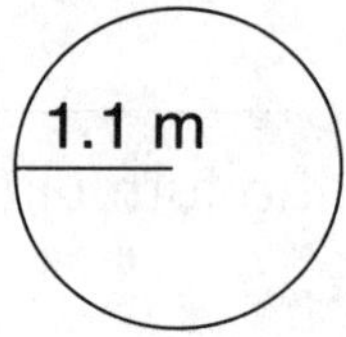

Which kind of house weighs the least?

A ____ ____ ____ ____ ____ ____ ____ ____ ____ ____
   200.96 32.15 176.62 3.8 50.24 3.8 63.58 314 78.5 28.26

44

Holt Mathematics

CHAPTER 9-6

Practice A
Area of Irregular Figures

Estimate the area of each figure. Each square represents 1 square foot. Choose the letter for the best answer.

1.

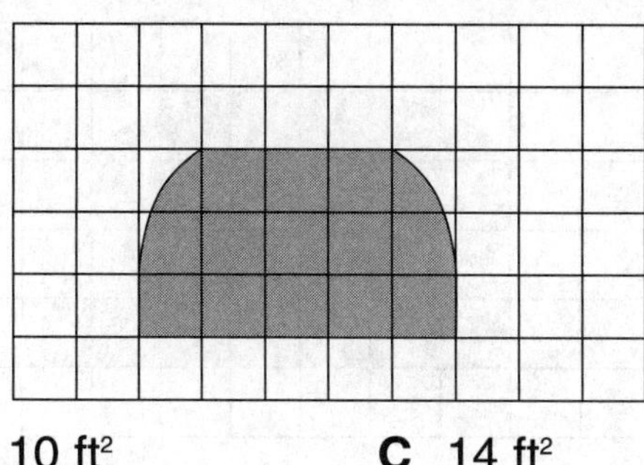

A 10 ft² **C** 14 ft²
B 11 ft² **D** 15 ft²

2.

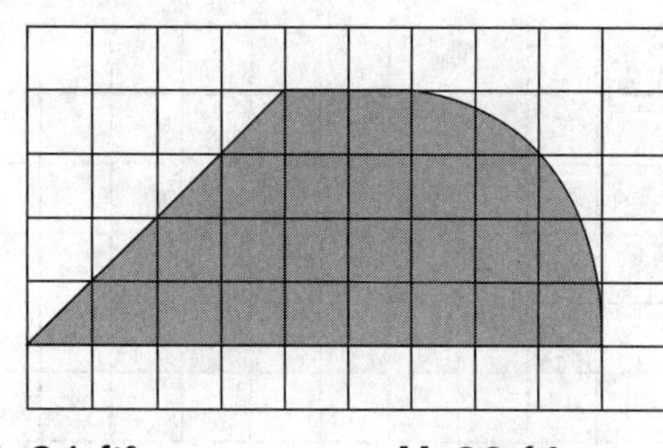

F 24 ft² **H** 32 ft²
G 27 ft² **J** 36 ft²

Find the area of each figure. Use 3.14 for π.

3.

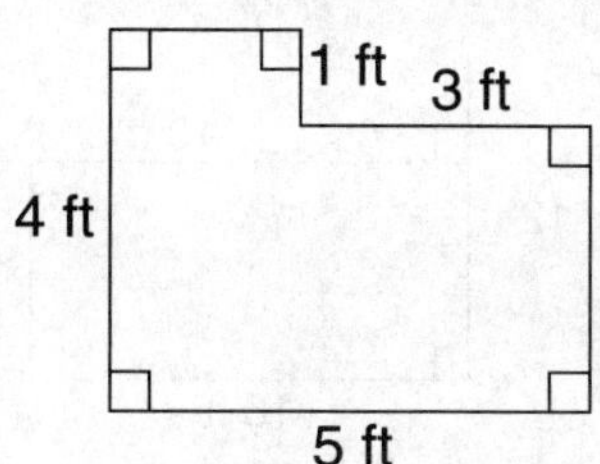

4.

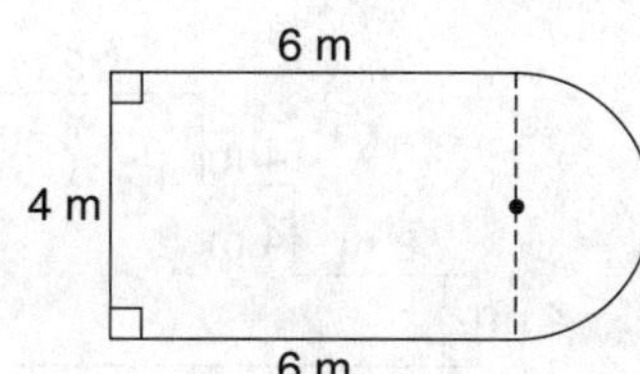

5.

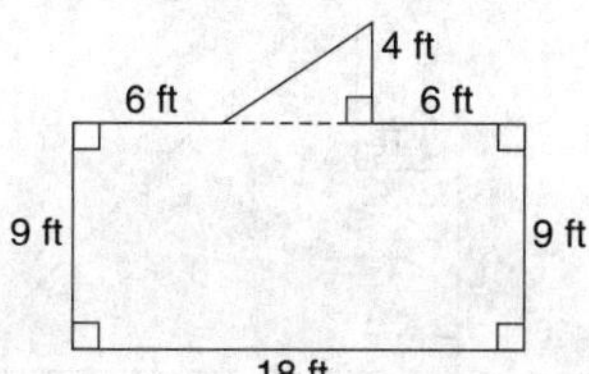

6.

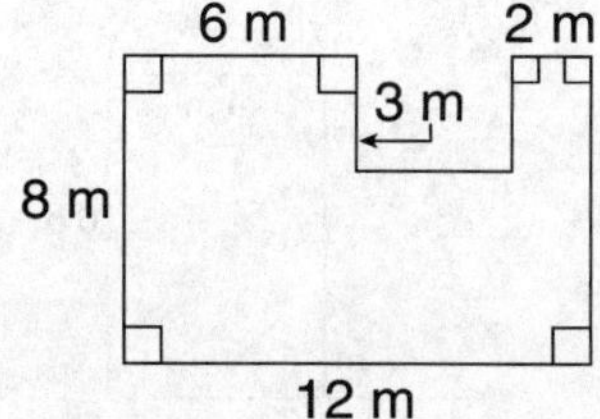

7.

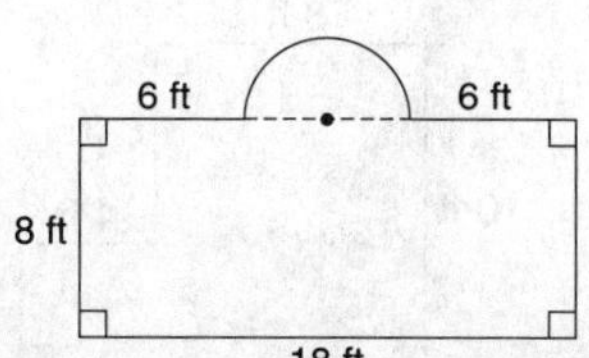

8.

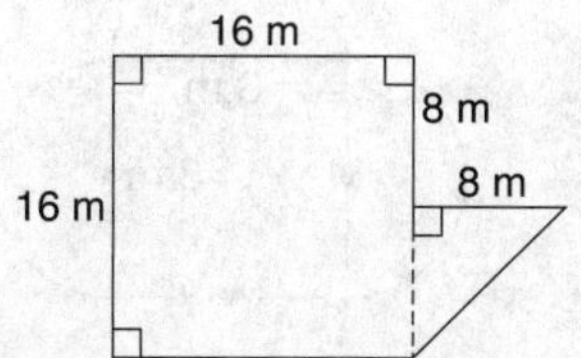

9. The figure shows the dimensions of a room. How much carpet is needed to cover the floor?

____

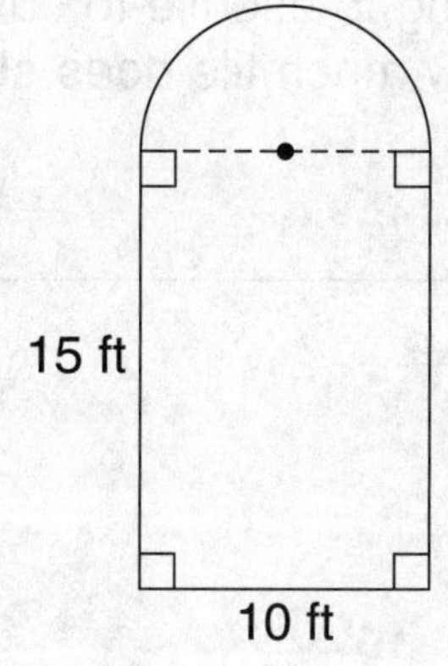

Holt Mathematics

Practice B
Area of Irregular Figures

Estimate the area of each figure. Each square represents 1 square foot.

1. 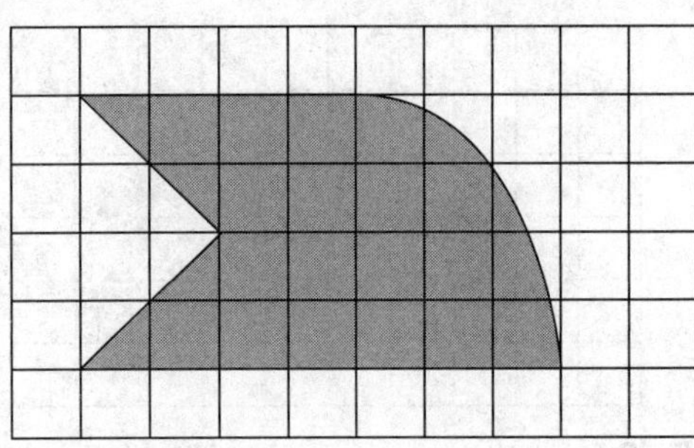

2. 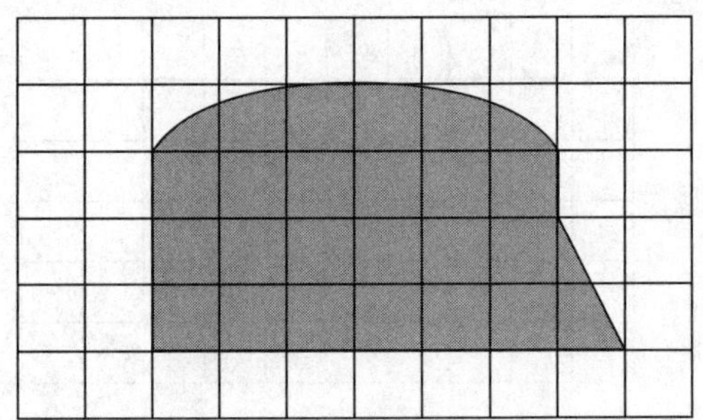

_______________________________   _______________________________

Find the area of each figure. Use 3.14 for π.

3. 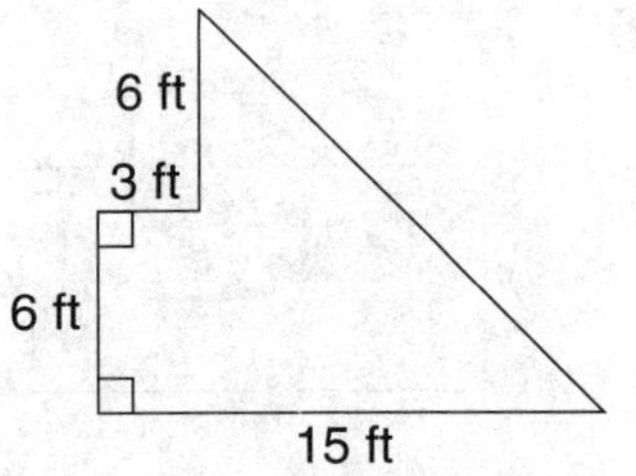

4. 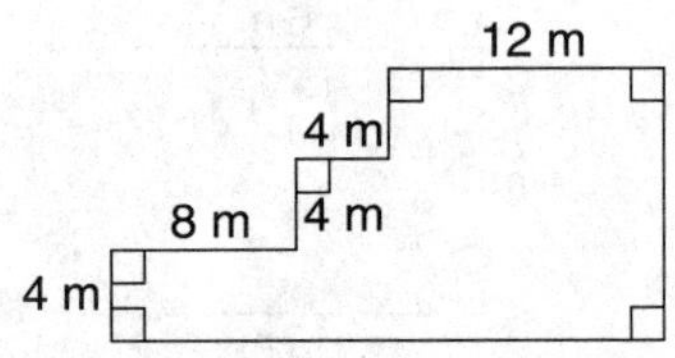

5. 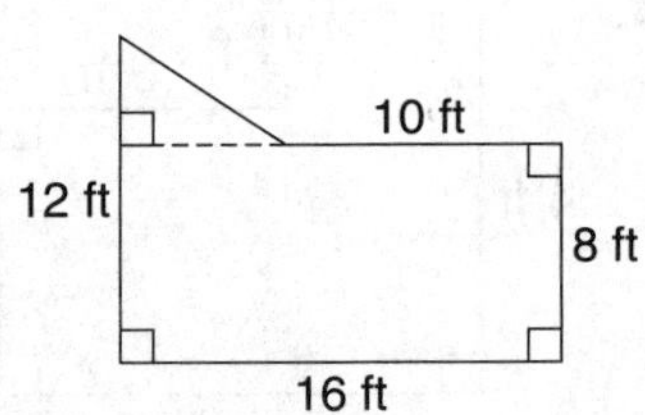

6. 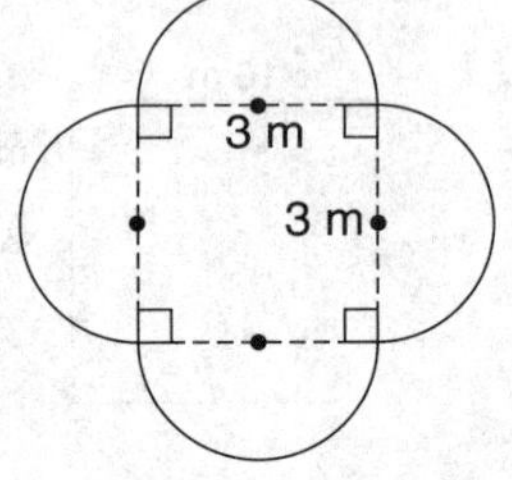

7. 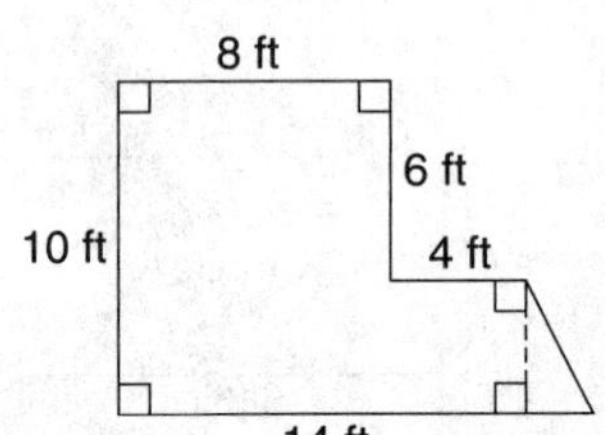

8. 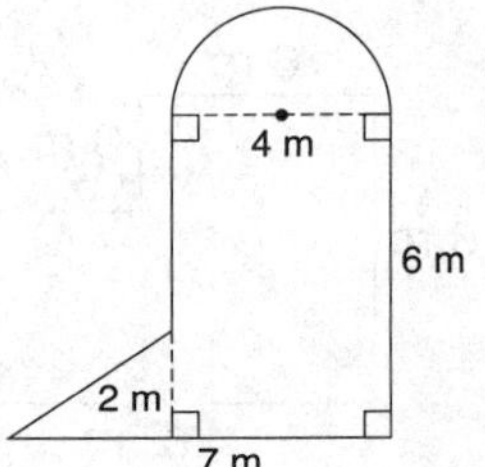

9. Marci is going to use tile to cover her terrace. How much tile does she need?

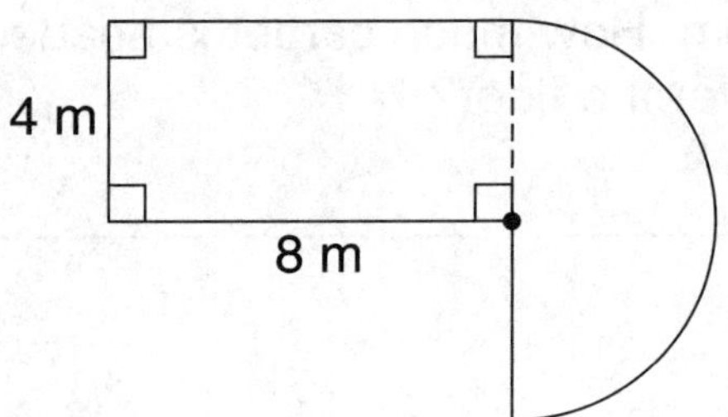

____

Holt Mathematics

Name _______________________________________ Date __________ Class __________

Practice C
Area of Irregular Figures

**Estimate the area of each figure. Each square represents
1 square foot.**

1.

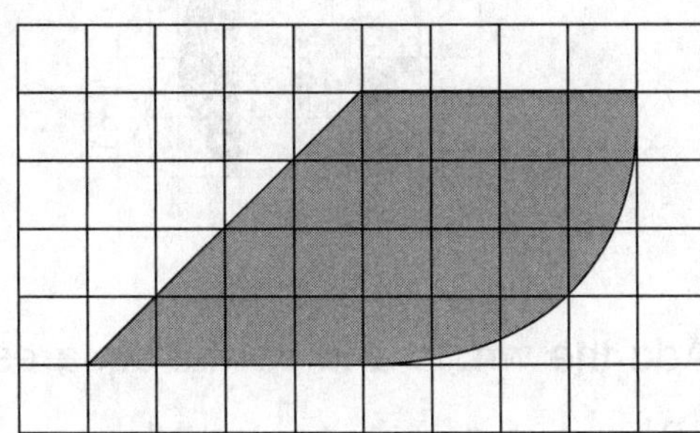

2.

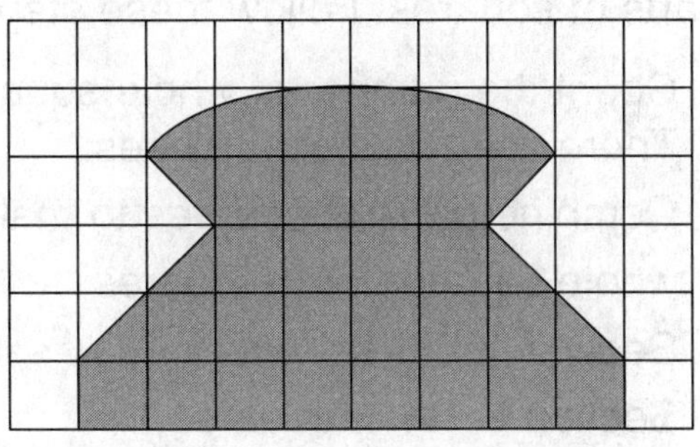

Find the area of each figure. Use 3.14 for π.

3.

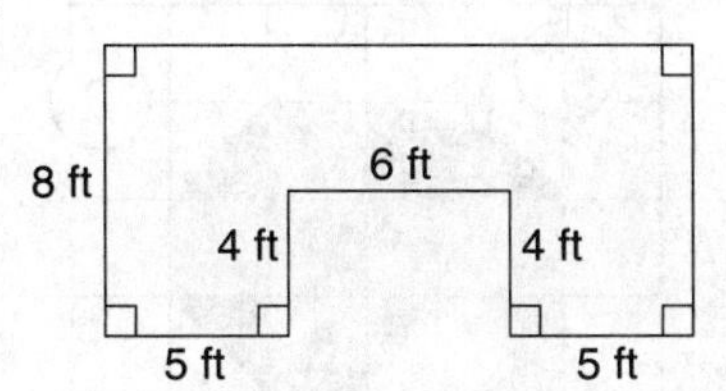

4.

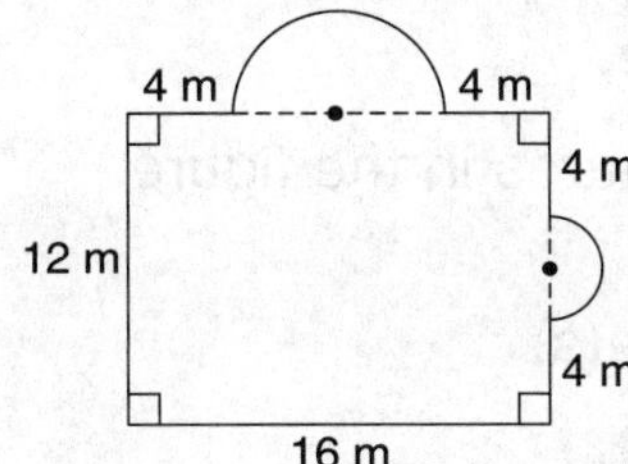

5.

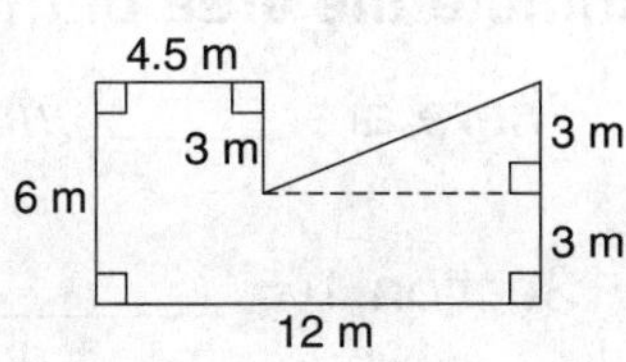

6. The figure shows the dimensions of a
room in which wedding receptions
are held. The room is being
carpeted. The three semi-circular
parts of the room are congruent. How
much carpet is needed?

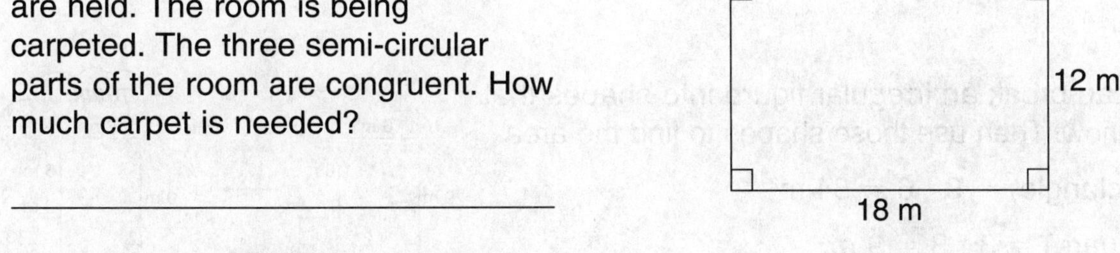

7. A polygon has vertices at $F(-5, 2)$,
$G(-3, 2)$, $H(-3, 4)$, $J(1, 4)$, $K(1, 1)$,
$L(4, 1)$, $M(4, -2)$, $N(6, -2)$, $P(6, -3)$,
and $Q(5, -3)$. Graph the figure on the
coordinate plane. Then find the area
and the perimeter of the figure.

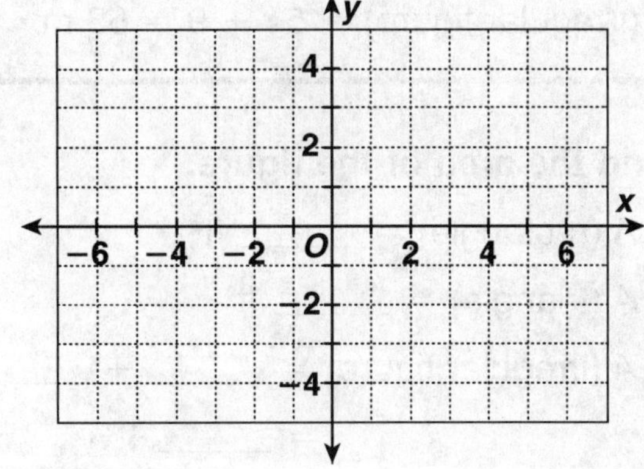

Holt Mathematics

Reteach
Area of Irregular Figures

When an irregular figure is on graph paper, you can estimate its area by counting whole squares and parts of squares. Follow these steps.

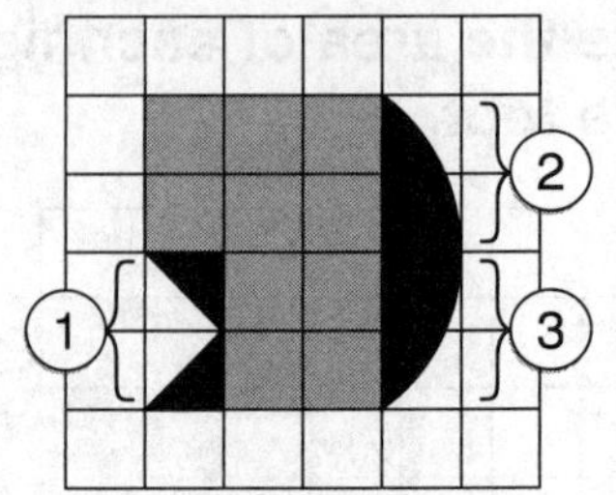

- Count the number of whole squares. There are 10 whole squares.
- Combine parts of squares to make whole squares or $\frac{1}{2}$-squares

  Section 1 = 1 square

  Section 2 $\approx 1\frac{1}{2}$ squares

  Section 3 $\approx 1\frac{1}{2}$ squares

- Add the whole and partial squares.

  $10 + 1 + 1\frac{1}{2} + 1\frac{1}{2} = 14$

  The area is about 14 square units.

Estimate the area of the figure.

1. There are _______ whole squares in the figure.

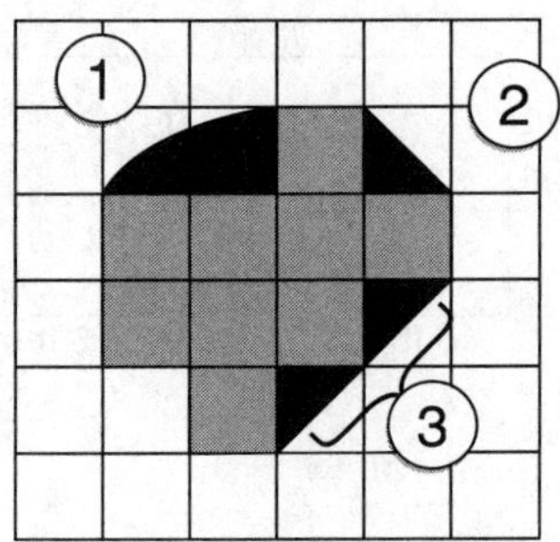

Section 1 $\approx$ _______ square(s)

Section 2 = _______ square(s)

Section 3 = _______ square(s)

A = _______ + _______ + _______ + _______ = _______ square units

You can break an irregular figure into shapes that you know. Then use those shapes to find the area.

A (rectangle) $= 9 \cdot 6 = 54$ m²

A (square) $= 3 \cdot 3 = 9$ m²

A (irregular figure) $= 54 + 9 = 63$ m²

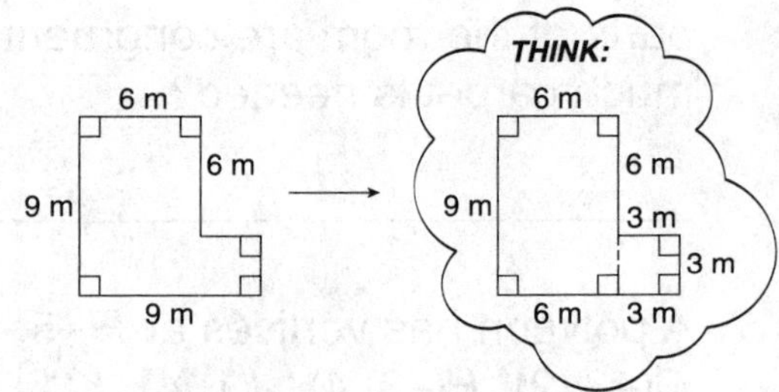

Find the area of the figure.

2. A (rectangle) = _______ ft²

A (triangle) = _______ ft²

A (irregular figure) = _______ + _______

= _______ ft²

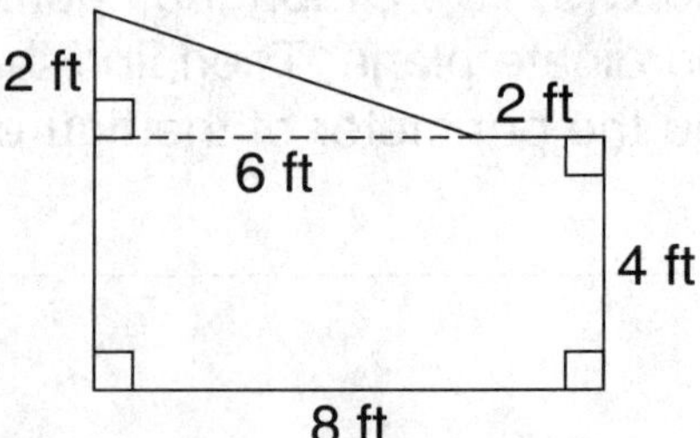

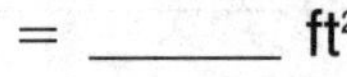

Holt Mathematics

CHAPTER 9-6

Challenge
Figure it Out!

Sometimes there is more than one way
to find the area of an irregular figure.

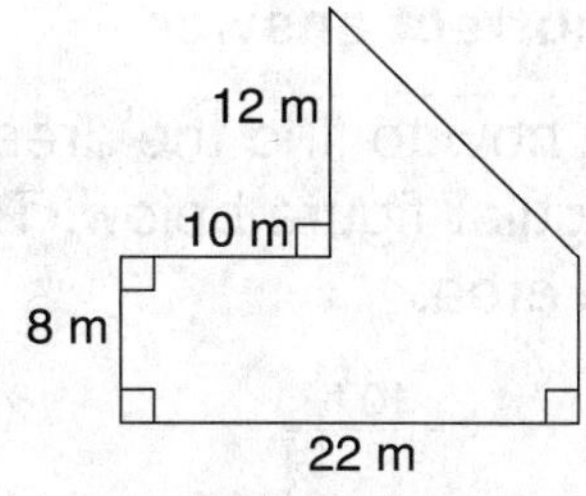

One Way

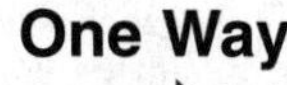
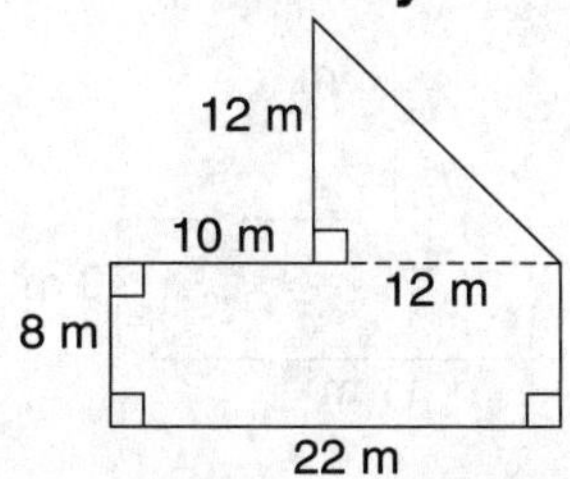

A(rectangle) = 22 · 8 = 176 ft²

A(triangle) = $\frac{1}{2}$ · 12 · 12 = 72 ft²

A(figure) = 176 + 72 = 248 ft²

Another Way

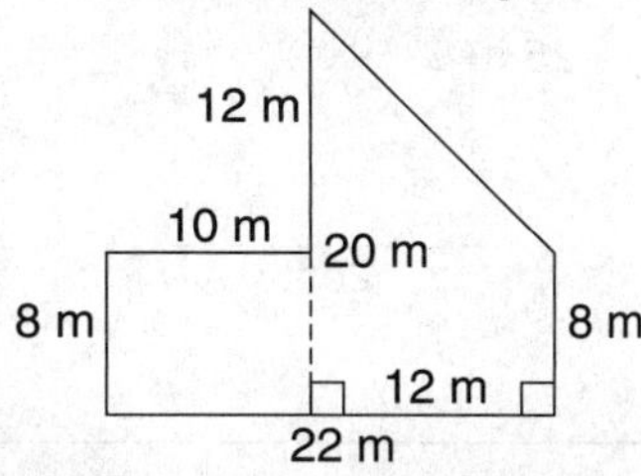

A(rectangle) = 10 · 8 = 80 ft²

A(trapezoid) = $\frac{1}{2}$ · 12 · (20 + 8) = 168 ft²

A(figure) = 80 + 168 = 248 ft²

Show two different ways to find the area of each figure.

	One Way	**Another Way**

1.

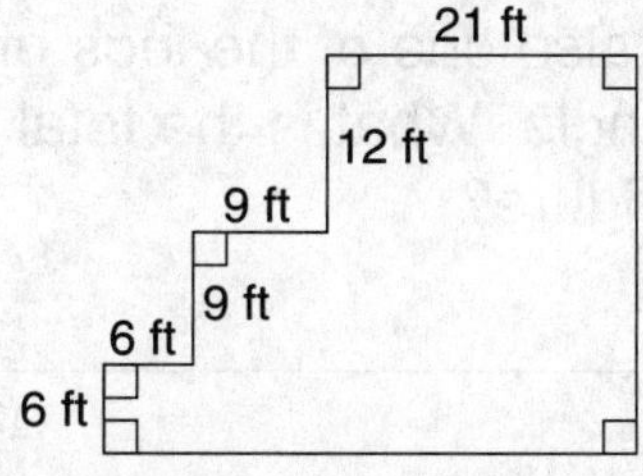

Area = __________   Area = __________

2.

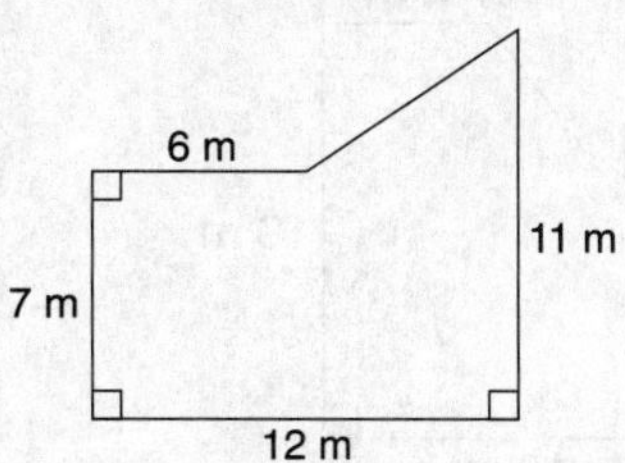

Area = __________   Area = __________

3.

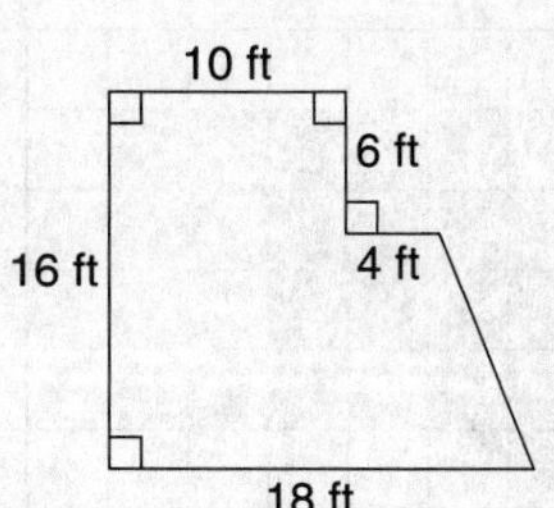

Area = __________   Area = __________

Holt Mathematics

Problem Solving
Area of Irregular Figures

Write the correct answer.

1. Explain how to find the area of the irregular figure below. Then find the area.

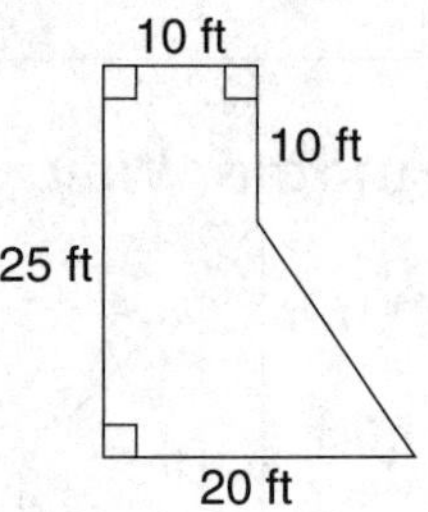

____

____

____

2. Mr. Bemis carpets the living room shown below. If he pays $20 per square meter, what is the total cost of the carpet?

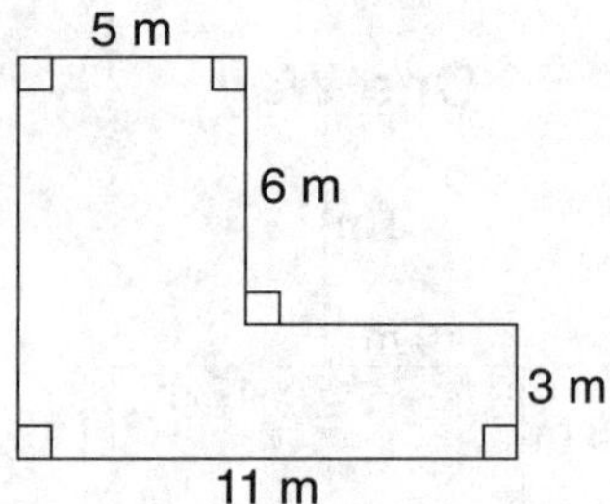

____

____

3. A figure is made of a square and a semi-circle. The square has sides of 16 cm each. One side of the square is also the diameter of the semi-circle. What is the total area of the figure?

____

4. A figure is made of a rectangle and an isosceles right triangle. The rectangle has sides of 6 in. and 3 in. One of the short sides of the rectangle is also one of the legs of the right triangle. What is the total area of the figure?

____

Choose the letter of the correct answer.

5. Norene builds the deck at the right. The area of the deck is 10 m² greater than was originally planned. What is the area of the deck?

A 110 m²   **C** 66 m²

B 76 m²   **D** 56 m²

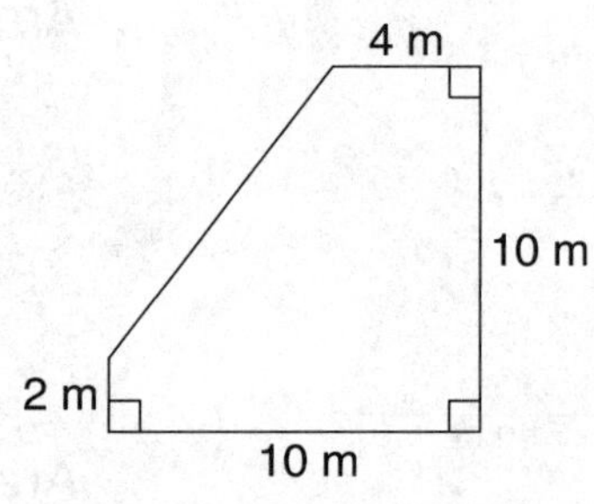

6. The grid to the right shows a swimming pool. Each square represents 1 square meter. What is the best estimate of the area of the swimming pool?

F 45 m²   **H** 37 m²

G 41 m²   **J** 32 m²

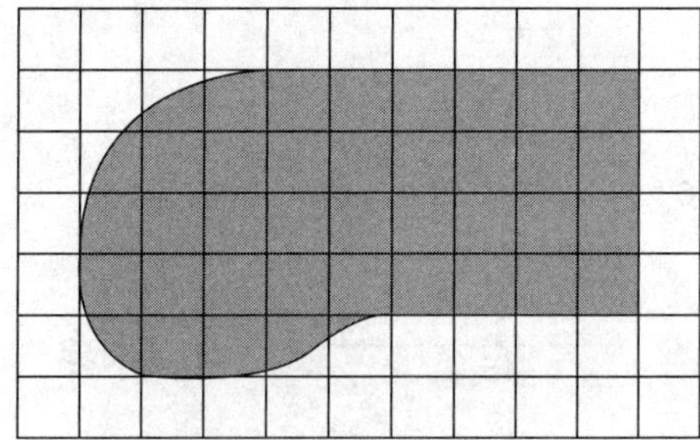

Holt Mathematics

Reading Strategies
Draw a Picture

Sometimes an irregular figure is made of shapes that you know,
such as rectangles, triangles, or semi-circles.

You can draw a line to break an irregular figure into other shapes.
Then you can use those shapes to help you find the area.

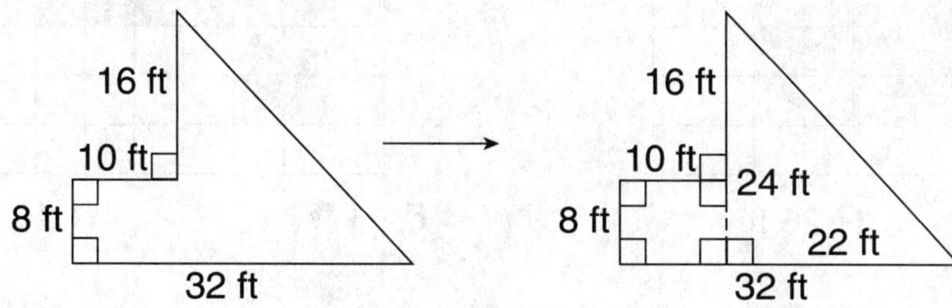

Answer each question.

1. A dashed line has been drawn through the figure. What two
   shapes does the dashed line create?

____

2. How can you find the area of the rectangle?

____

3. What is the area of the rectangle?

____

4. How can you find the area of the right triangle?

____

5. What is the area of the right triangle?

____

6. How can you use the area of the rectangle and the area of the
   right triangle to find the total area of the irregular figure?

____

7. What is the total area of the irregular figure?

____

Holt Mathematics

Puzzles, Twisters, & Teasers

CHAPTER 9-6 *What's Bugging You?*

Each square represents 1 square foot. Estimate the area of each figure. Circle the letter next to the better estimate.

1.

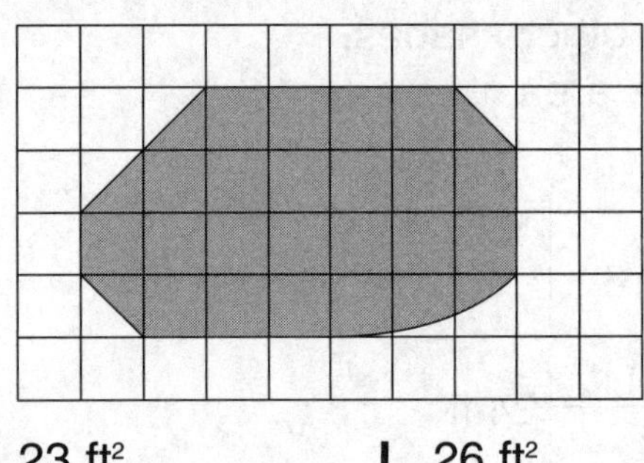

D 23 ft²   **I** 26 ft²

2.

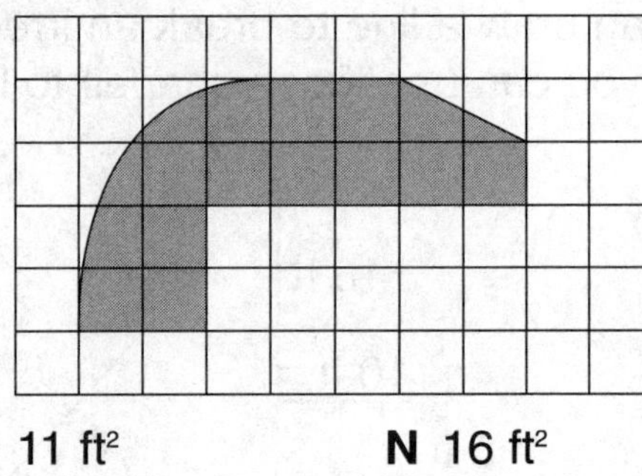

F 11 ft²   **N** 16 ft²

Next, find the area of each figure. Circle the letter next to your answer.

3.

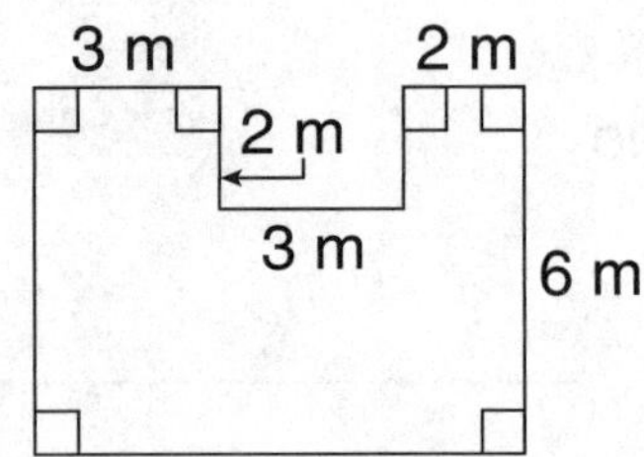

L 38 m²
R 42 m²

4.

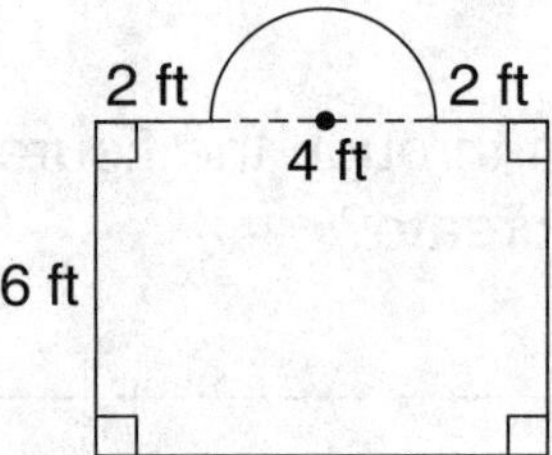

A 54.28 ft²
E 60.56 ft²

5.

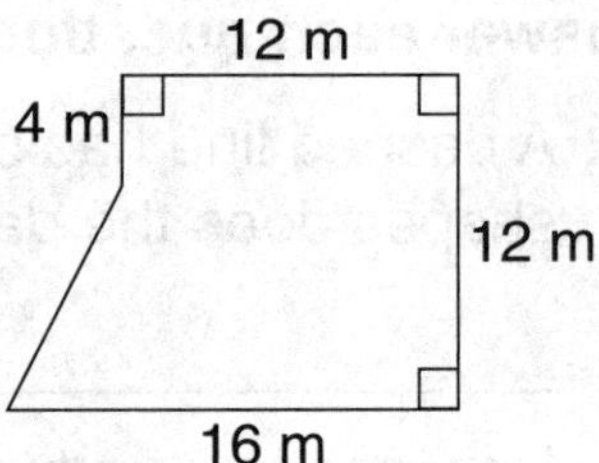

O 160 m²
E 168 m²

6.

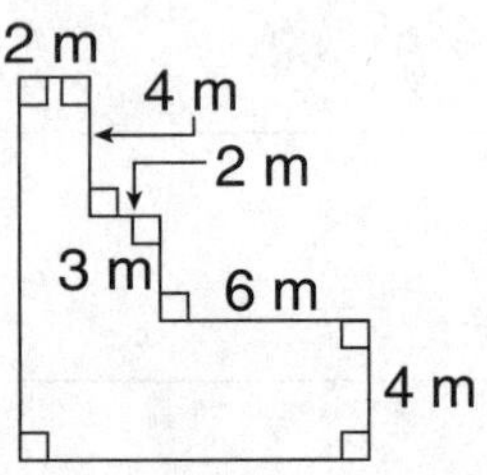

H 52 m²
L 60 m²

7.

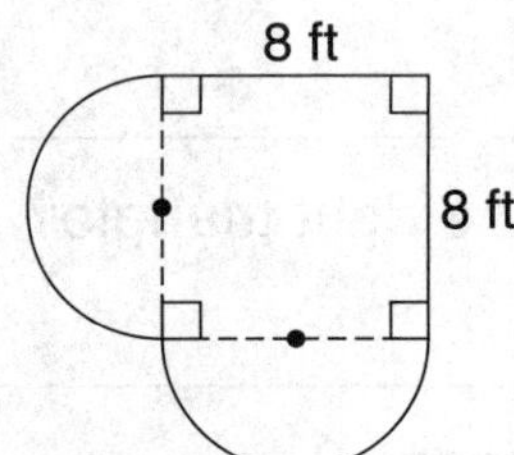

S 89.12 ft²
Y 114.24 ft²

8.

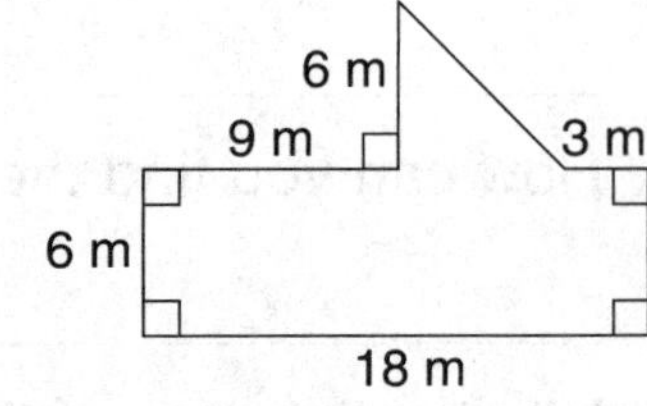

F 126 m²
B 144 m²

Write the circled letters above the problem numbers to solve the riddle.

What kind of insect breathes fire?

A ___ ___ ___   G ___ ___ ___ ___ ___
  1.   3.   4.   5.   2.   8.   6.   7.

Holt Mathematics

LESSON 9-7

Practice A
Powers and Roots

Find each square.

1. 5^2 _________

2. 10^2 _________

3. 11^2 _________

4. 13^2 _________

5. 7^2 _________

6. 12^2 _________

Find each square root.

7. $\sqrt{225}$ _________

8. $\sqrt{36}$ _________

9. $\sqrt{81}$ _________

10. $\sqrt{289}$ _________

11. $\sqrt{196}$ _________

12. $\sqrt{484}$ _________

Match each square root in Column A with the letter of its nearest whole number estimate in Column B. Use a calculator to check your answer.

Column A	Column B	Column A	Column B
13. $\sqrt{38}$ _____	**A.** 2	**19.** $\sqrt{84}$ _____	**G.** 10
14. $\sqrt{19}$ _____	**B.** 8	**20.** $\sqrt{46}$ _____	**H.** 4
15. $\sqrt{66}$ _____	**C.** 6	**21.** $\sqrt{95}$ _____	**I.** 8
16. $\sqrt{54}$ _____	**D.** 5	**22.** $\sqrt{10}$ _____	**J.** 7
17. $\sqrt{29}$ _____	**E.** 7	**23.** $\sqrt{15}$ _____	**K.** 9
18. $\sqrt{5}$ _____	**F.** 4	**24.** $\sqrt{71}$ _____	**L.** 3

25. The area of a checkerboard is 112 in^2. What is the approximate length of each side of the checkerboard? Find your answer to the nearest inch.

____

Holt Mathematics

LESSON 9-7 · Practice B
Powers and Roots

Find each square.

1. 6^2 _______ **2.** 19^2 _______ **3.** 13^2 _______ **4.** 4^2 _______

5. 10^2 _______ **6.** 14^2 _______ **7.** 20^2 _______ **8.** 18^2 _______

Find each square root.

9. $\sqrt{289}$ _______ **10.** $\sqrt{49}$ _______ **11.** $\sqrt{256}$ _______ **12.** $\sqrt{81}$ _______

13. $\sqrt{121}$ _______ **14.** $\sqrt{625}$ _______ **15.** $\sqrt{576}$ _______ **16.** $\sqrt{900}$ _______

17. $\sqrt{11}$ _______ **18.** $\sqrt{31}$ _______ **19.** $\sqrt{98}$ _______ **20.** $\sqrt{50}$ _______

21. $\sqrt{152}$ _______ **22.** $\sqrt{14}$ _______ **23.** $\sqrt{70}$ _______ **24.** $\sqrt{28}$ _______

25. $\sqrt{39}$ _______ **26.** $\sqrt{193}$ _______ **27.** $\sqrt{119}$ _______ **28.** $\sqrt{85}$ _______

29. $\sqrt{5}$ _______ **30.** $\sqrt{42}$ _______ **31.** $\sqrt{75}$ _______ **32.** $\sqrt{215}$ _______

33. The area of a square vegetable garden is 75 ft^2. What is the approximate length of each side of the garden? Find your answer to the nearest foot. _______________

34. The area of a computer screen is 138 in^2. What is the approximate length of each side of the screen? Find your answer to the nearest inch. _______________

35. Tim broke a square picture window with his baseball. The area of the window is 52 ft^2. What is the approximate width of the window to be replaced? Find your answer to the nearest foot. _______________

36. A square tile has an area of 413 cm^2. What is the approximate length of a side of the tile? Find your answer to the nearest centimeter. _______________

Holt Mathematics

Practice C

LESSON 9-7

Powers and Roots

Find each square

1. 18^2 _______

2. $(0.7)^2$ _______

3. 27^2 _______

4. $(3.9)^2$ _______

5. 50^2 _______

6. $(0.31)^2$ _______

7. $\left(\dfrac{1}{5}\right)^2$ _______

8. $\left(\dfrac{7}{8}\right)^2$ _______

Find each square root.

9. $\sqrt{784}$ _______

10. $\sqrt{484}$ _______

11. $\sqrt{225}$ _______

12. $\sqrt{961}$ _______

Estimate each square root to the nearest whole number. Use a calculator to check your answer.

13. $\sqrt{54}$ _______

14. $\sqrt{86}$ _______

15. $\sqrt{12}$ _______

16. $\sqrt{27}$ _______

17. $\sqrt{71}$ _______

18. $\sqrt{15}$ _______

19. $\sqrt{95}$ _______

20. $\sqrt{118}$ _______

21. $\sqrt{149}$ _______

22. $\sqrt{201}$ _______

23. $\sqrt{250}$ _______

24. $\sqrt{175}$ _______

25. $\sqrt{450}$ _______

26. $\sqrt{375}$ _______

27. $\sqrt{600}$ _______

28. $\sqrt{1,000}$ _______

Estimate each sum or difference to the nearest whole number.

29. $\sqrt{20} + \sqrt{30}$ _______

30. $\sqrt{50} + \sqrt{10}$ _______

31. $\sqrt{75} - \sqrt{60}$ _______

32. $\sqrt{7} + 7^2$ _______

33. $5^2 - \sqrt{5}$ _______

34. $\sqrt{13} + 13^2$ _______

35. Find the area of a square whose perimeter is 48 in. _______________

36. Find the perimeter of a square whose area is 225 yd^2. _______________

37. The area of a square rug is 3,000 in^2. What is the
approximate length of each side of the rug?
Find your answer to the nearest inch. _______________

Holt Mathematics

Reteach
Powers and Roots

Recall that the formula for the area of a square is $A = s^2$. A power in which the exponent is 2 is called a *square*.

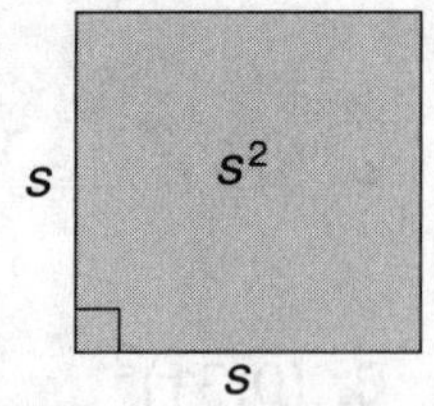

Use the model to help you find the square.

1. 5^2

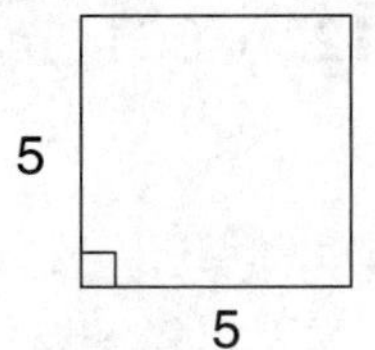

$A = s^2$

$A = $ ______^{2}

$A = $ ______ • ______

$A = $ ______

2. 9^2

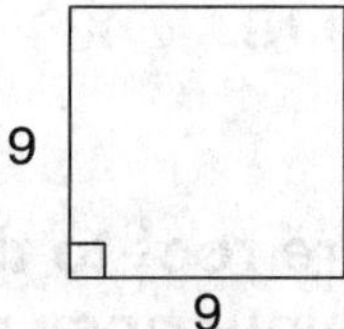

$A = s^2$

$A = ($______$)^2$

$A = $ ______ • ______

$A = $ ______

Find the square.

3. 13^2 **4.** 16^2 **5.** 11^2 **6.** 7^2

_______   _______   _______   _______

The numbers 36 and 81 are *perfect squares.* **Perfect squares** are numbers that are the squares of whole numbers.

The square root of 36 is 6. $\sqrt{36} = 6$ because $6 \cdot 6 = 36$.

A table of square roots can help you estimate the square root of a number that is not a perfect square.

Square Root	1	2	3	4	5	6	7	8	9	10
Perfect Squares	1	4	9	16	25	36	49	64	81	100

7. $\sqrt{44}$

Find the perfect square nearest 44.

44 is closest to ______.

Since $\sqrt{49} = $ ______,

$\sqrt{44}$ is closest to ______.

8. $\sqrt{87}$

Find the perfect square nearest 87.

87 is closest to ______.

Since $\sqrt{81} = $ ______,

$\sqrt{87}$ is closest to ______

Holt Mathematics

LESSON	**Challenge**
9-7	*Power Patterns*

1. Complete the tables below.

Powers of 3		Powers of 7		Powers of 9	
3^1	3	7^1	7	9^1	
3^2		7^2		9^2	
3^3		7^3		9^3	
3^4		7^4		9^4	
3^5		7^5		9^5	
3^6		7^6		9^6	
3^7		7^7		9^7	
3^8		7^8		9^8	

2. Describe the pattern for the ones digits of the powers of 3 values.

____

3. What is the ones digit of 3^9? of 3^{31}? of 3^{98}? ___________________

4. Describe the pattern of the ones digits of the powers of 7 values.

____

5. What is the ones digit of 7^{10}? of 7^{24}? of 7^{53}? ___________________

6. Do the ones digits for the values of the powers of all whole numbers follow a pattern of *four* repeating digits? Explain.

____

7. What is the ones digit of 11^3? of 11^4? of 11^5? of 11^6? ___________________

8. The ones digits of the values of the powers of 11 are the same as the ones digits of the powers of what number? ___________________

9. Describe the patterns shown by the powers of 5.

____

10. Describe the pattern of the ones digits for the powers of 6.

____

Holt Mathematics

LESSON
9-7

Problem Solving
Powers and Roots

Write the correct answer. For Problems 1 and 2, use the following formula to find the distance in meters a free-falling object falls from a place of rest: $d = 0.5 \cdot 9.8 \cdot t^2$ (t = time in seconds).

1. As part of a science experiment, Hsing drops a ball from the roof of the school. How far does the ball fall in 2 seconds?

2. Mel drops a stone from the edge of a cliff overlooking the ocean. How far does the stone fall in 5 seconds?

3. At the county fair, the apple pies are lined up side-by-side for judging on a 6-foot table. Each pie has an area of 50.24 in^2. How many pies are on the table?

4. The community swimming pool has an area of 1,024 square feet. The pool is in the shape of a square. What is the perimeter of the pool?

Choose the letter of the correct answer.

5. The Portuguese national flag is a rectangle. In the center of the flag is a coat of arms and shield on a circle. This circle has a diameter that is half the flag's height. If the flag's circle has an area of 3.14 square feet, what is the height of the flag?

   A 1 ft

   B 3 ft

   C 4 ft

   D 2 ft

6. A square picture has an area of 81 square inches. The perimeter of the frame for the picture is 8 inches longer than the perimeter of the picture itself. What is the length of each side of the square frame for this picture?

   F 8 in.

   G 10 in.

   H 9 in.

   J 11 in.

7. A basketball game starts with a jump ball. This occurs in the center of the basketball court within a circle that has an area of 113.04 square feet. What is the radius of this circle?

   A 3 ft

   B 6 ft

   C 9 ft

   D 36 ft

8. A square fence encloses a vegetable garden with an area of 169 square feet. What is the perimeter of the fence?

   F 52 ft

   G 26 ft

   H 13 ft

   J 56.25 ft

Holt Mathematics

Reading Strategies

LESSON 9-7

Use A Graphic Organizer

A **square number** is a whole number multiplied by itself.

$3 \times 3 = 9$ **9** is a square number.

$6 \times 6 = 36$ **36** is a square number.

A square number has two factors that are the same.

A **square root** is the opposite of a square number.

This chart helps you picture square numbers and square roots.

Perfect Square	Not a Perfect Square
4 by 4 $A = \ell w$ (4 • 4) $A = 4^2 \rightarrow$ Read: "four squared." $A = 16$ square units 16 is a square number.	4 by 5 $A = \ell w$ (4 • 5) $A = 20$ $A = 20$ square units 20 is not a square number.

Squares and Square Roots

Square Root

- The opposite of a square
- $\sqrt{}$ is the radical symbol used to write square roots.

$\sqrt{16} = 4$ $\sqrt{20} \approx 4.47$

Read: "The square root of 16 equals 4."   Read: "The square root of 20 equals approximately 4.47."

Use the chart to help you answer each question.

1. How can you tell if a number is a square number?

____

2. Write how you would read 5^2.

____

3. Complete. $5^2 = $ _______________________________

4. Write how you would read. $\sqrt{25}$.

____

Holt Mathematics

Puzzles, Twisters & Teasers

LESSON 9-7 *The Root of the Matter!*

**Find and circle the words in the box below in the word search.
The words may be horizontal or vertical. Find a word in the word
search that solves the riddle. Circle it and write it on the line.**

power	root	perfect	square	radical
sign	express	value	symbol	model

```
P E R F E C T R C V B
O V U J E R V A L U E
W G T Y U I L D S Q W
E X P R E S S I Y E R
R M J I O Q I C M Z X
O V B N V U G A B O I
O I O P L A N L O P U
T B H V I R U S L L I
S A M O D E L A S E C
```

Why did the computer sneeze?

It had a ____________________________ .

60

Holt Mathematics

Practice A
The Pythagorean Theorem

**Use the Pythagorean Theorem to find each missing measure.
Choose the letter for the best answer.**

1.

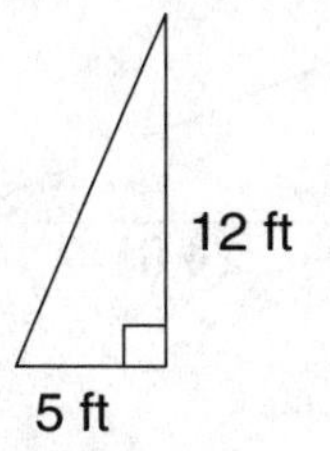

A 7 ft	**C** 14 ft
B 13 ft	**D** 17 ft

2.

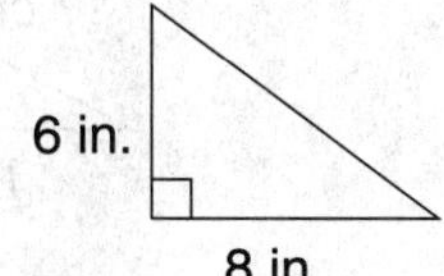

F 14 in.	**H** 5 in.
G 10 in.	**J** 2 in.

3.

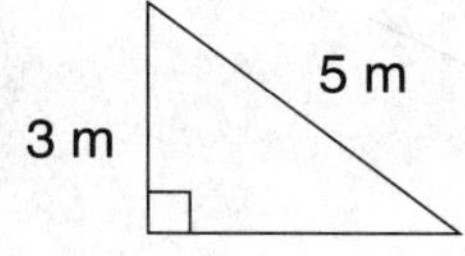

A 2 m	**C** 6 m
B 4 m	**D** 8 m

4.

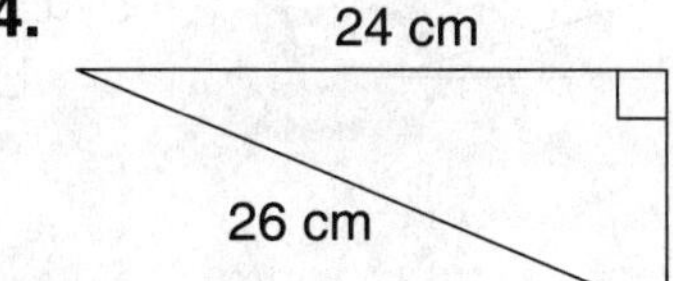

F 2 cm	**H** 10 cm
G 4 cm	**J** 40 cm

5.

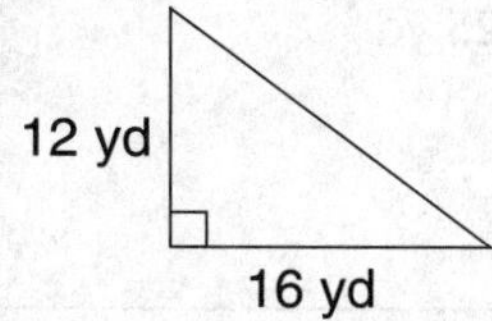

A 11 yd	**C** 20 yd
B 16 yd	**D** 28 yd

6.

F 23 in.	**H** 13 in.
G 17 in.	**J** 7 in.

7. A 15-foot ladder is leaning against a
wall. The ladder is 6 ft from the base
of the wall. About how far above the
ground does the ladder touch the
wall? Round your answer to the
nearest tenth.

____

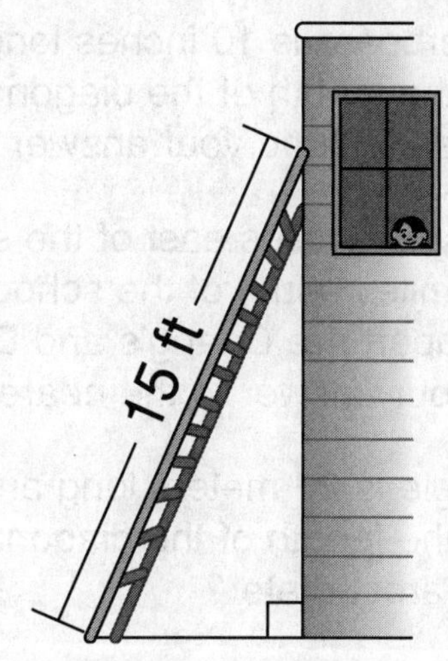

Holt Mathematics

Practice B
The Pythagorean Theorem

Use the Pythagorean Theorem to find each missing measure.

1. 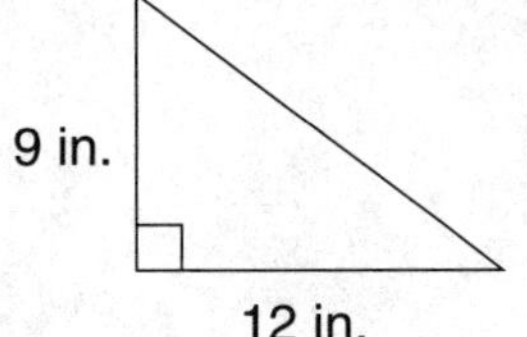

2. 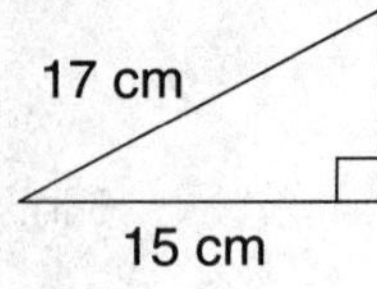

3. 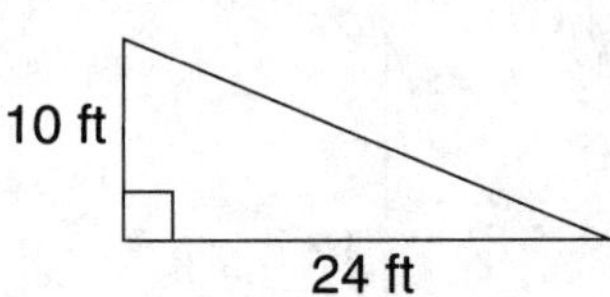

____

4. 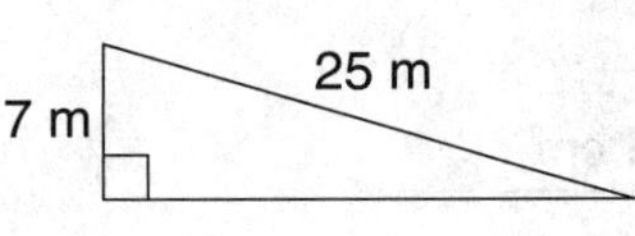

5. 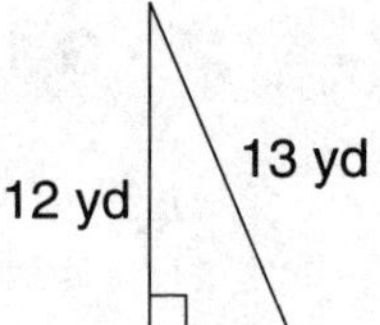

6. 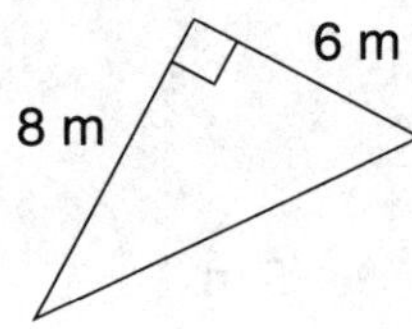

____

7. 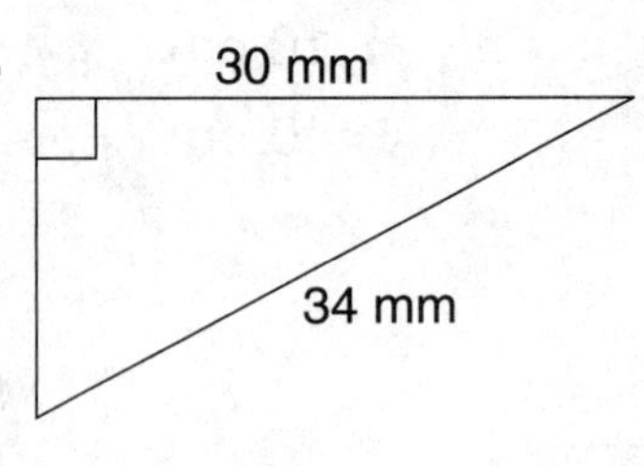

8. 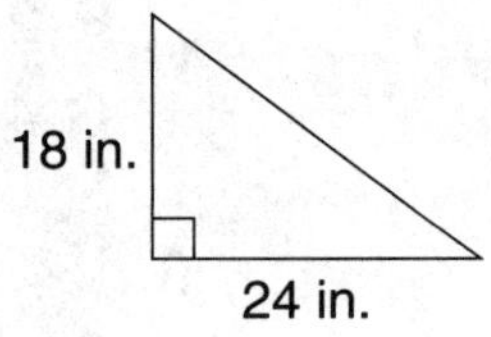

9. 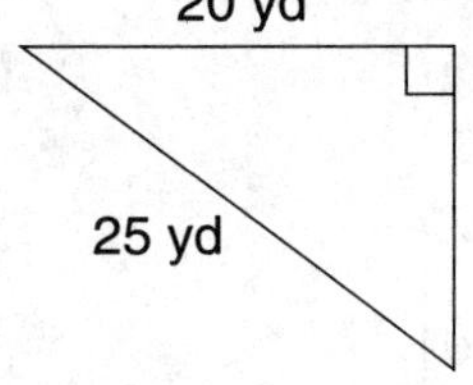

____

10. A 20-ft ladder is leaning against a wall. If the ladder is 12 ft from the base of the wall, how high above the ground does the ladder touch the wall? _______________

11. A checkerboard is 10 inches long on each side. What is the length of the diagonal from one corner to another? Round your answer to the nearest tenth. _______________

12. Chang lives 8 miles east of the school. Deborah lives 12 miles south of the school. Approximately how far apart are Chang's and Deborah's homes? Round your answer to the nearest tenth. _______________

13. A rectangle is 26 meters long and 18 meters wide. What is the length of the diagonal of the rectangle to the nearest meter? _______________

Holt Mathematics

Practice C
The Pythagorean Theorem

Use the Pythagorean Theorem to find each missing measure.

1. 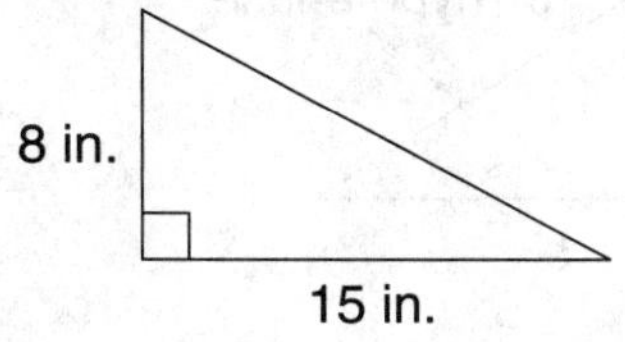

____

2. 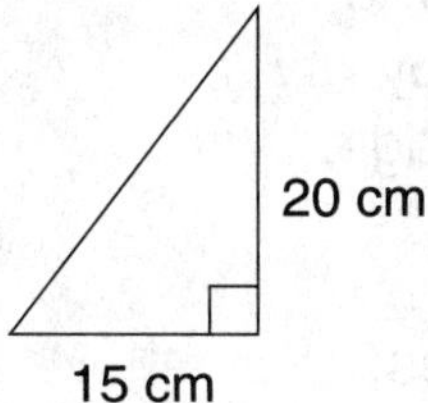

____

3. 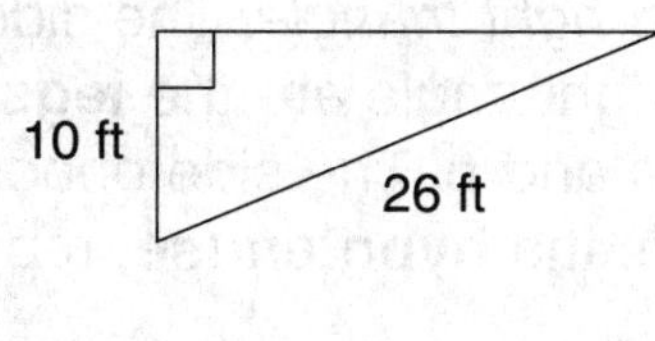

____

4. 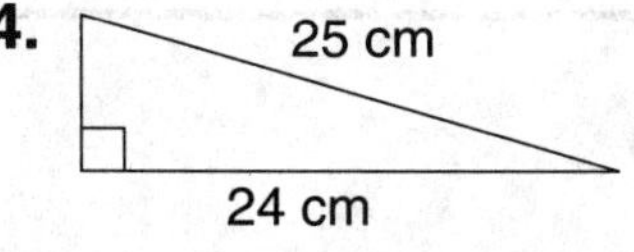

____

5. 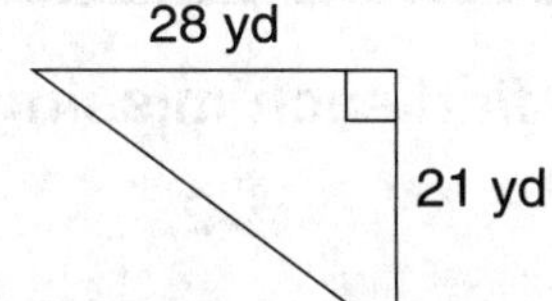

____

6. 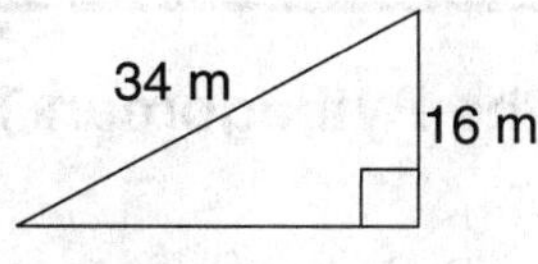

____

**The lengths of two sides of a right triangle are given. Find the
length of the third side to the nearest tenth.**

7. legs: 4 ft and 9 ft

8. leg: 8 m; hypotenuse: 12 m

____

9. leg: 15 ft; hypotenuse: 20 ft

10. legs: 10 m and 7 m

____

Determine whether each set of lengths forms a right triangle.

11. $a = 10$, $b = 24$, $c = 25$ _______

12. $a = 12$, $b = 16$, $c = 20$ _______

13. $a = 4$, $b = 6$, $c = 8$ _______

14. $a = 5$, $b = 12$, $c = 13$ _______

Solve.

15. The foot of a ladder is 8 ft from the base of a wall. If the
ladder is 20 ft long, how high up on the wall does the
ladder reach? Round your answer to the nearest tenth. _________

16. The hypotenuse of an isosceles right triangle
is 16 centimeters. Find the length of the two
other sides of the triangle to the nearest tenth. _________

Holt Mathematics

<table><tr><td>**LESSON**
9-8</td><td># Reteach
The Pythagorean Theorem</td></tr></table>

A triangle containing a right angle is called
a *right triangle*. The sides adjacent to the
right angle are the **legs,** represented by
a and *b*. The side opposite the right angle
is the **hypotenuse,** represented by *c*.

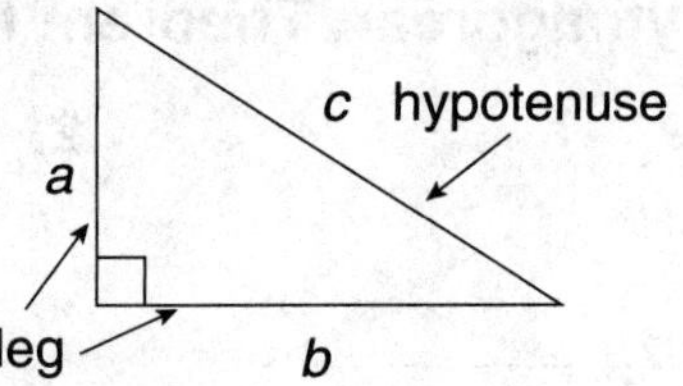

The **Pythagorean Theorem** states:

If *a* and *b* are the lengths of the legs of a right triangle
and *c* is the length of the hypotenuse, then $a^2 + b^2 = c^2$.

Use the Pythagorean Theorem to find each missing length.

1.

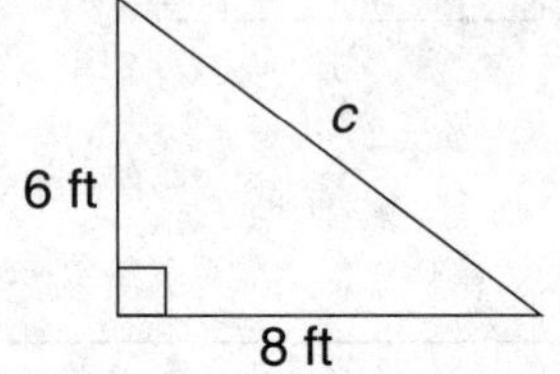

$$a^2 + b^2 = c^2$$

$$\underline{\quad}^2 + \underline{\quad}^2 = c^2$$

$$\underline{\quad} + \underline{\quad} = c^2$$

$$\underline{\qquad} = c^2$$

$$c = \underline{\qquad}$$

2.

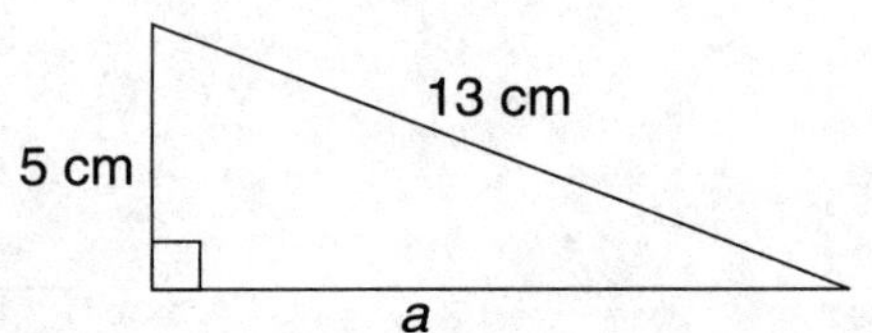

$$a^2 + b^2 = c^2$$

$$a^2 + \underline{\quad}^2 = \underline{\quad}^2$$

$$a^2 + \underline{\quad} = \underline{\qquad}$$

$$a^2 = \underline{\qquad}$$

$$a = \underline{\qquad}$$

3.

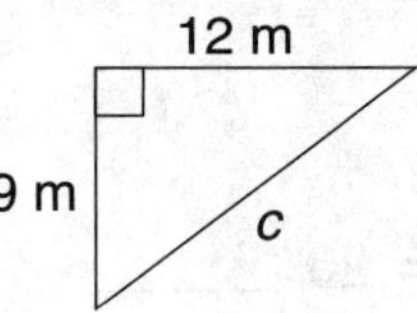

4.

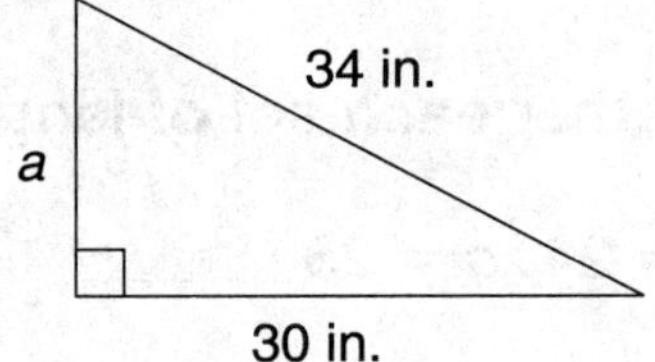

5.

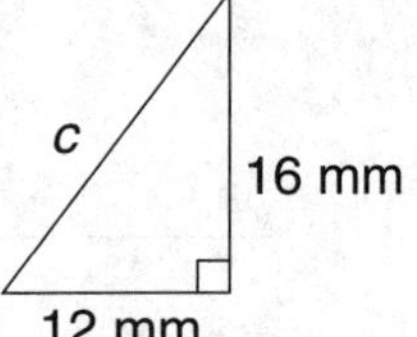

6.

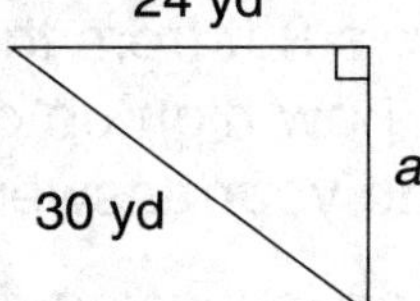

Holt Mathematics

LESSON 9-8 · Challenge
Triangle Families

Any three numbers that satisfy the formula $a^2 + b^2 = c^2$ form a *Pythagorean triple*.

$3^2 + 4^2 = 5^2$

$9 + 16 = 25$

$(3, 4, 5)$ is a Pythagorean triple.

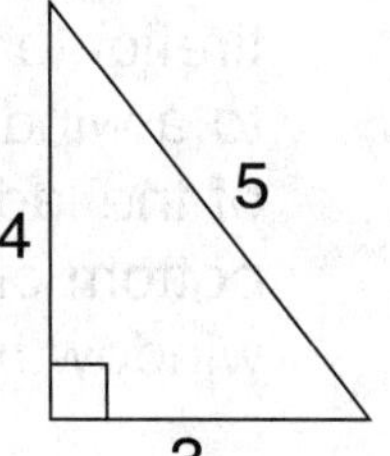

When you multiply or divide a Pythagorean triple by the same number, you get another Pythagorean triple.

So, $(3 \cdot 2, 4 \cdot 2, 5 \cdot 2)$,

or $(6, 8, 10)$,

is a Pythagorean triple.

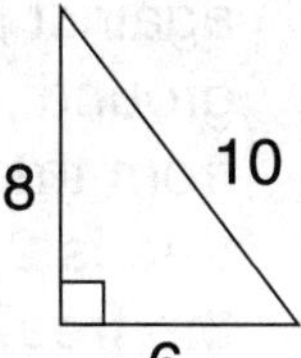

Check it out: $6^2 + 8^2 = 10^2$

$36 + 64 = 100$

Complete each table to find new Pythagorean triples. Then check some of your results.

1. The (3, 4, 5) Family

Multiply by 3.	(9, 12, 15)
Multiply by 5.	
Multiply by 10.	
Divide by 5.	

2. The (5, 12, 13) Family

Multiply by 2.	
Multiply by 3.	
Multiply by 8.	
Divide by 10.	

3. The (7, 24, 25) Family

Divide by 4.	
Multiply by 2.	
Multiply by 4.	
Multiply by 12.	

4. The (8, 15, 17) Family

Divide by 2.	
Multiply by 3.	
Multiply by 4.	
Multiply by 20.	

5. Use the formula to check one triple of the (3, 4, 5) family.

____

6. Use the formula to check one triple of the (5, 12, 13) family.

____

7. Use the formula to check one triple of the (7, 24, 25) family.

____

Holt Mathematics

LESSON 9-8 Problem Solving
The Pythagorean Theorem

Write the correct answer.

1. During a storm, a tree falls toward a house. The top of the tree leans against the house 45 feet above the ground. The distance on the ground from the house to the base of the tree is 24 feet. What is the height of the tree?

2. During a training exercise, a firefighter leans a 40-foot ladder up to a window in a house. The bottom of the ladder is 24 feet from the bottom of the house. How high is the window from the ground?

3. A triangle has a hypotenuse of 25 centimeters and a base of 20 centimeters. What is the area of this right triangle?

4. The football field at the University of Texas at Arlington is 60 yards by 100 yards. Is the length of the diagonal across this field more or less than 200 yards? Explain.

Choose the letter of the correct answer.

5. The minimum size of a soccer field for players under 8 years of age is 20 yards by 30 yards. About how far is the diagonal distance on a field with these dimensions?

   A about 12 yd   **C** about 36 yd

   B about 25 yd   **D** about 45 yd

6. The minimum size of a soccer field for international matches is 70 yards by 110 yards. If a player runs diagonally across this field, about how much farther does she run than if the field were 50 yards by 100 yards?

   F about 242 yd   **H** about 112 yd

   G about 130 yd   **J** about 19 yd

7. In the state of Virginia, Winchester is 21 miles north of Front Royal. Arlington is 58 miles east of Front Royal. What is the distance to the nearest mile from Winchester to Arlington?

   A 37 mi   **C** 89 mi

   B 62 mi   **D** 441 mi

8. On a child's slide, the distance from the bottom rung to the top of the ladder is 6 feet. The straight distance from the bottom rung of the ladder to the bottom of the slide is 36 inches. About how long is the slide?

   F about 6.7 ft   **H** about 9.0 ft

   G about 8.4 ft   **J** about 36.5 ft

Holt Mathematics

Reading Strategies

LESSON 9-8

Understanding Vocabulary

A **right triangle** has special features.

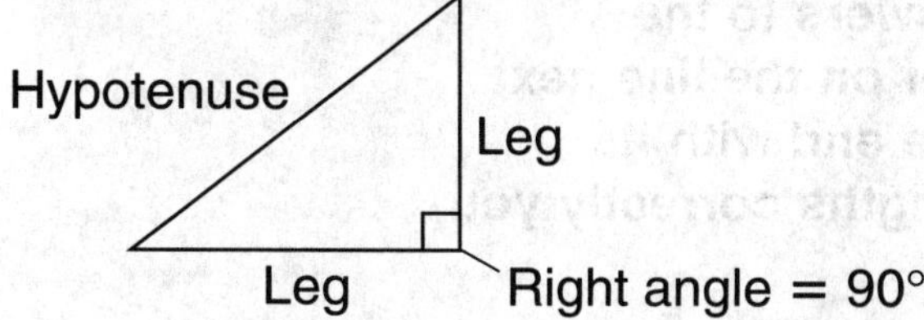

Use the triangle to complete each sentence.

1. A right triangle has one right angle that measures _________________ degrees.

2. The two shorter sides of a right angle are called _________ _________.

3. The longest side of a right triangle is called the _________________.

The **Pythagorean Theorem** describes a special relationship among the sides of a right triangle. The theorem states that the sum of the legs squared is equal to the hypotenuse squared.

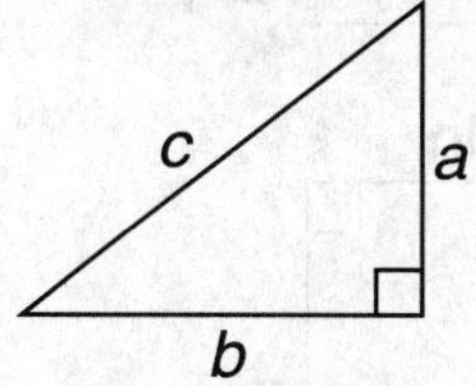

$$a^2 + b^2 = c^2$$

Look at the figure of the triangle to answer each question.

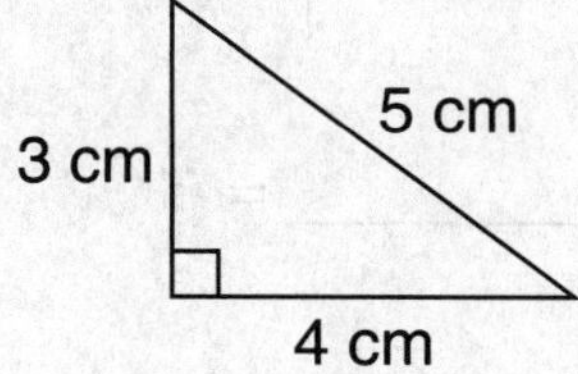

4. What is 3 cm squared? _________________

5. What is 4 cm squared? _________________

6. What is the total of the squares of the legs? _________________

7. What is 5 cm squared? _________________

8. What do you notice about the sum of the squares of the legs and the square of the hypotenuse? _________________

Holt Mathematics

Puzzles, Twisters & Teasers

LESSON 9-8

Fair and Square!

Use the Pythagorean Theorem to find the length of the hypotenuse of each triangle. Round your answers to the nearest tenth, if necessary. Write your answer on the line next to the letter. Replace the answers listed at the end with its corresponding letter. If you calculated the lengths correctly you will solve the riddle!

1. __________ = K

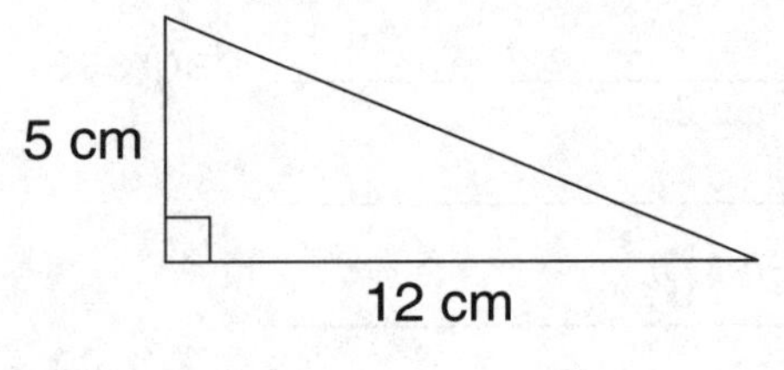

2. __________ = E

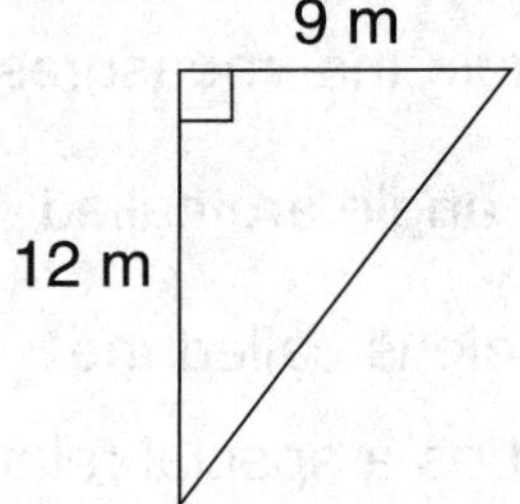

3. __________ = N

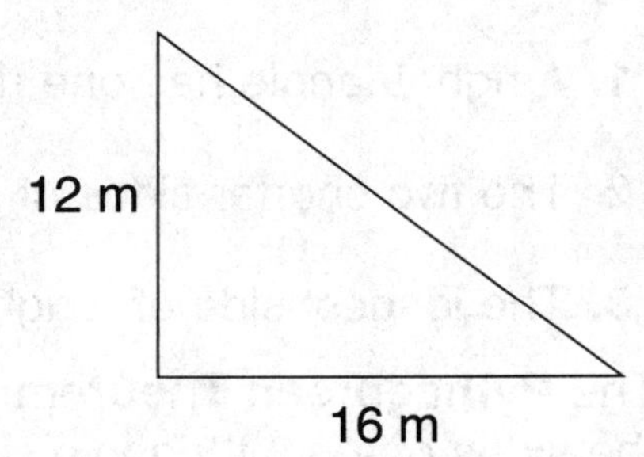

4. __________ = B

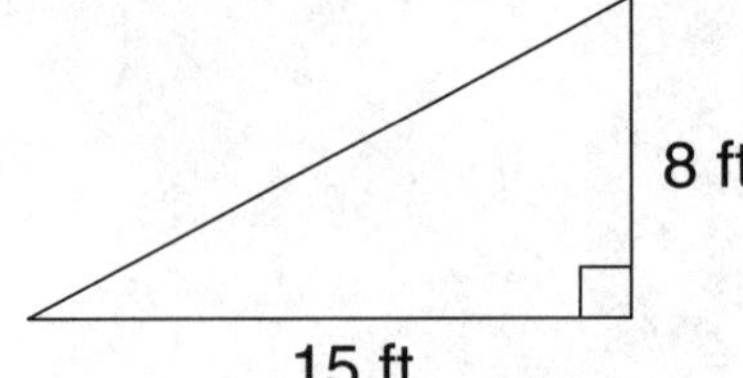

5. __________ = U

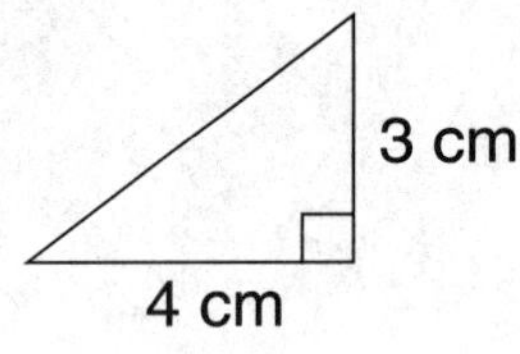

6. __________ = I

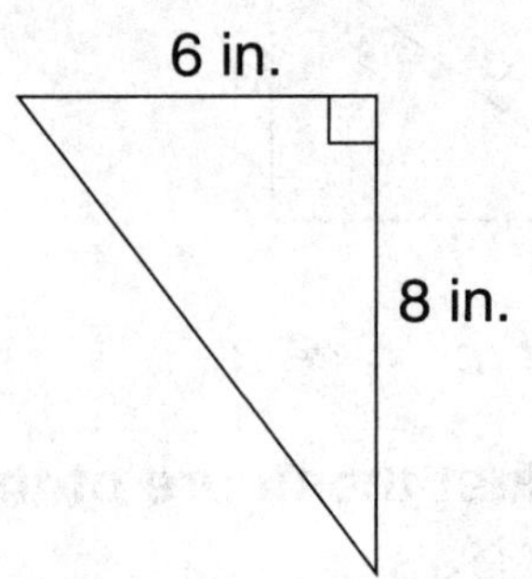

7. __________ = R

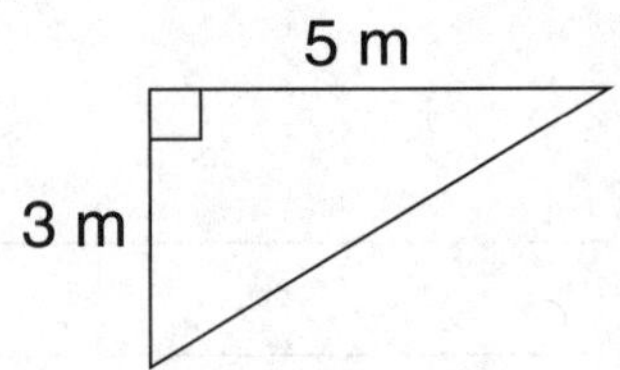

8. __________ = G

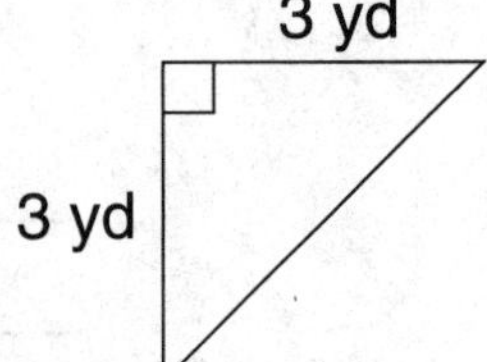

9. __________ = T

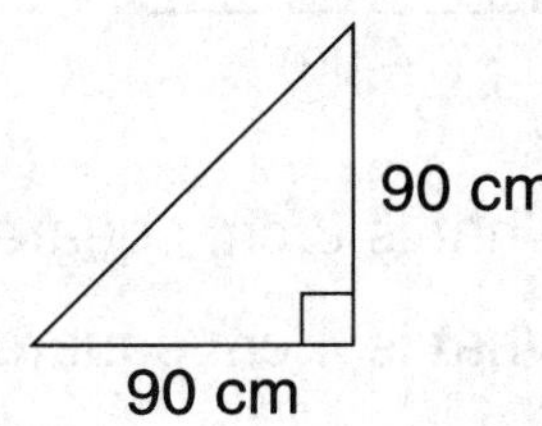

Why did the woman want an elephant instead of a car?

It had a ____ ____ ____ ____ ____ ____ ____ ____ ____ ____ ____ .
     17   10   4.2   4.2   15   5.8   127.3 5.8   5   20   13

Holt Mathematics

LESSON 9-1
Practice A
Accuracy and Precision

Choose the more precise measurement in each pair.

1. 5 ft, 53 in.   2. 10 qt, 44 c   3. 853 g, 1 kg

    53 in.   **44 c**   **853 g**

Determine the number of significant digits in each measurement. Choose the letter for the best answer.

4. 3,000
- (A) 1
- B 2
- C 3
- D 4

5. 1.50
- F 1
- G 2
- (H) 3
- J 4

6. 3.405
- A 1
- B 2
- C 3
- (D) 4

7. 0.0006
- (F) 1
- G 4
- H 5
- J 6

8. 201
- A 2
- (B) 3
- C 4
- D 5

9. 50.020
- F 2
- G 3
- H 4
- (J) 5

Calculate. Find each answer with the correct number of significant digits in the box. Each answer can be used only once.

| 13.6 | 4.8 | 30 | 5 | 66 | 28 | 7.6 | 8 | 28.2 | 66.4 | 4.77 | 14 |

10. 19 − 5.4 = **14**   11. 23.6 + 6 = **30**   12. 11.8 − 7 = **5**

13. 12.3 • 5.4 = **66**   14. 6.0 • 4.7 = **28**   15. 30.5 ÷ 4.0 = **7.6**

16. 0.329 • 14.49 = **4.77**   17. 58.3 + 8.12 = **66.4**   18. 59.04 ÷ 2.09 = **28.25**

19. 32.6 − 27.83 = **4.8**   20. 9 − 1.4 = **8**   21. 12.9 + 0.71 = **13.6**

3
Holt Mathematics

LESSON 9-1
Practice B
Accuracy and Precision

Choose the more precise measurement in each pair.

1. 2 tons, 3,700 lb   2. 4 weeks, 27 days   3. 3.5 m, 3.03 m

   3,700 lb   **27 days**   **3.03 m**

4. 3 ft, 32 in.   5. 4.6 mL, 2.8 L   6. 15.8 km, 15 km

   32 in.   **4.6 mL**   **15.8 km**

Determine the number of significant digits in each measurement.

7. 5.801 **4**   8. 0.06 **1**   9. 75,000 **2**

10. 0.00007 **1**   11. 100,000,000 **1**   12. 300.080 **6**

13. 9.007 **4**   14. 0.840 **3**   15. 0.0050 **2**

Calculate. Use the correct number of significant digits in each answer.

16. 21 − 8.6 = **12**   17. 47.6 + 8 = **56**   18. 9.8 − 3 = **7**

19. 31.3 − 24.78 = **6.5**   20. 9.63 + 3.4 = **13.0**   21. 15.7 + 0.82 = **16.5**

22. 0.54 + 0.104 = **0.64**   23. 102 − 2.77 = **99**   24. 62 + 0.319 = **62**

25. 52.7 • 2.3 = **120**   26. 8.0 • 1.7 = **14**   27. 20.5 ÷ 6.0 = **3.4**

28. 23.9 • 14.4 = **344**   29. 19.2 ÷ 0.03 = **600**   30. 1,240 ÷ 4.025 = **308**

31. 0.18 • 6.2 = **1.1**   32. 95 ÷ 32 = **3.0**   33. 74.3 • 0.22 = **16**

4
Holt Mathematics

LESSON 9-1
Practice C
Accuracy and Precision

Choose the more precise measurement in each pair.

1. 6 L, 5,320 mL   2. 27 oz, 2 lb   3. 400 min, 6.5 hr

   5,320 mL   **27 oz**   **400 min**

4. 5 ft, 60 in.   5. 8.7 km, 870 m   6. 8 cm, 2.5 cm

   60 in.   **870 m**   **2.5 cm**

Determine the number of significant digits in each measurement.

7. 2.0801 **5**   8. 0.03 **1**   9. 10,500 **3**

10. 0.00307 **3**   11. 900,000,000 **1**   12. 302.80 **5**

13. 7.007 **4**   14. 0.840 **3**   15. 0.080 **2**

Calculate. Use the correct number of significant digits in each answer.

16. 170 − 6.8 = **160**   17. 37.6 + 8 = **46**   18. 2.8 ÷ 0.3 = **9**

19. 51.37 − 24.7 = **26.7**   20. 6.89 + 7.3 = **14.2**   21. 95.7 + 9.893 = **105.6**

22. 72.5 • 3.2 = **230**   23. 2,600 • 4.71 = **12,000**   24. 25 ÷ 7.04 = **3.6**

Which unit is more precise?

25. quart, gallon **quart**   26. ounce, pound **ounce**

27. pint, cup **cup**   28. day, month **day**

29. gram, kilogram **gram**   30. inch, foot **inch**

5
Holt Mathematics

LESSON 9-1
Reteach
Accuracy and Precision

A measurement is more **precise** than another measurement if its unit of measure is smaller.

1. Which measurement is more precise, 14 cm or 140 mm?
   a. 14 cm means the measure is to the nearest **centimeter**.
   b. 140 mm means the measure is to the nearest **millimeter**.
   c. Since a millimeter is a smaller measurement, **140 mm** is more precise.

2. Which measurement is more precise, 5 ft or 50.1 ft?
   a. 5 ft means the measure is to the nearest **foot**.
   b. 50.1 ft means the measure is to the nearest tenth of a **foot**.
   c. Since a tenth of a foot is a smaller measurement, **50.1 ft** is more precise.

Choose the more precise measurement in each pair.

3. 7.5 m or 75 cm   4. 11.0 in. or 11 in.   5. 8.4 lb or 8 oz

   75 cm   **11.0 in.**   **8 oz**

You can find the number of **significant digits**, or all the digits that are known to be exact, by dividing a measurement by its smallest place value. The number of digits in the quotient is the number of significant digits.

Measurement	Smallest Place Value	Measurement ÷ Smallest Place Value	Number of Significant Digits
14 ft	1	14 ÷ 1 = 14	2
0.043 in.	0.001	0.043 ÷ 0.001 = 43	2
50.1 m	0.1	50.1 ÷ 0.1 = 501	3

Zeros to the left of a decimal point are not significant if there are no digits to the right of the decimal point.

   910 contains 2 significant digits: 9 and 1.

Zeros after digits to the right of a decimal point are significant.

   64.9500 contains 6 significant digits: 6, 4, 9, 5, and the two zeros.

Determine the number of significant digits.

6. 41.25 **4**   7. 30.6 **3**   8. 0.085 **2**

9. 1,207,000 **4**   10. 38.600 **5**   11. 4.00020 **6**

6
Holt Mathematics

Challenge
The Significance of a Riddle

Write a solution for each of the following riddles. Your answer must include all of the information in the riddle, but all of the information about each number is not always given in the riddle. **Possible answers are given.**

1. I am a number with *only* the digits 1, 2, and 3. I have four significant digits and no decimal places. What number could I be?

   1,231

2. I am a number with *only* the digits 2, 4, and 0. I have one decimal place and four significant digits. What number could I be?

   240.2

3. I am a number with *only* the digits 6 and 0. I have three significant digits and four decimal places. What number could I be?

   0.0660

4. I am the greatest number with *only* the digits 1, 2, and 3. I have five significant digits. What number am I?

   33,321

5. I have no significant digits. What number am I?

   0

6. What number could I be with two significant digits and five decimal places?

   0.00050

7. I am the least number with *only* the digits 1, 2, and 3, with five significant digits and six decimal places. What number am I?

   0.010123

8. I am a number with two significant digits and four decimal places. I begin and end with 0. What number could I be?

   0.0070

9. I am a number between 155 and 1,555. I have five significant digits and two decimal places. What number could I be?

   200.05

10. I am a number between 0 and 2 with *only* the digits 1 and 4. I have five significant digits and four decimal places. What number could I be?

    1.1414

11. I have four digits, but only three are significant. I have three decimal places. What number could I be?

    0.120

12. I have six digits, but only two are significant. I have no decimal places. What number could I be?

    450,000

7

Problem Solving
Accuracy and Precision

Write the correct answer.

1. Normal rainfall in Hilo, Hawaii, is 2.36 feet per year. Yearly rainfall in Honolulu, Hawaii, is 7.77 inches. Which measurement is more precise? Explain.

   An inch is a smaller unit than a foot, so 7.77 in. is more precise than 2.36 ft.

2. The Seismosaurus was about 5.5 meters high. The Troodon, considered the smartest dinosaur, was only about 1.75 meters high. Write the difference in height of the two dinosaurs using the correct number of significant digits.

   3.8 m difference

3. Esther drives a total of 120 miles to and from work each day. She works 5 days per week. How many significant digits are there in the number of miles Esther drives to and from work each week?

   1 significant digit

4. Big Al weighs 256.8 pounds. His brother, Little Lou, weighs 125 pounds. What is their combined weight? Express your answer with the appropriate number of significant digits.

   382 lb

Choose the letter of the correct answer.

5. A square picture frame measures 50.1 centimeters on each side. How many significant digits are there in the perimeter of the frame?

   A 2 (C) 4
   B 3 D 1

6. Four students each measured his or her own height. The measures below show each student's results. Which is the most precise measure?

   F 5 ft H 62 in.
   G 5.5 ft (J) 64.5 in.

7. Before the year 2000, Harvard University had 12,877,360 books in its library. How many significant digits are there in the number of books in the Harvard Library?

   A 5 (C) 7
   B 6 D 8

8. In 1996, Tokyo Disneyland had an estimated 16.98 million visitors. Disneyland in Anaheim, California, had 15 million visitors. What is the estimated total number of visitors for both parks using the correct number of significant digits?

   (F) 32 million H 31.9 million
   G 31.98 million J 31 million

8

Reading Strategies
Analyze Information

Precise means as close or accurate as possible. No measurement is absolutely exact. It is important when measuring to be as precise or exact as possible.
You can choose the more precise measurement.

The width of your little finger: 0.5 inch or 0.46 inch

Since 0.46 has a smaller place value than 0.5, it is the more precise measurement.

The width of a computer screen: 1 foot or 13 inches

Since inches are smaller than feet, 13 inches is the more precise measurement.

Complete each problem.

1. What does the word *precise* mean?

   as close or accurate as possible

2. Which measurement is more precise, inches or feet? Explain.

   Inches are more precise, since they are smaller units of measure.

3. Why is 0.46 inch more precise than 0.5 inch?

   because 0.46 has a smaller decimal place value

Circle the more precise measurement in each pair below.

4. $3\frac{1}{2}$ feet   (**$3\frac{1}{3}$ feet**)

5. (**34 mm**)   3 cm

6. (**7 feet**)   2 yards

7. $\frac{1}{2}$ hour   (**35 minutes**)

8. (**23 ounces**)   2 pounds

9. 1 hour   (**53 minutes**)

10. (**27 days**)   1 month

9

Puzzles, Twisters & Teasers
Those Darn Details!

Find and circle the words below in the word search. Words may be horizontal or vertical or backwards. Find a word in the word search to solve the riddle. Circle it and write it on the line.

precision   measurement   determine   level   accuracy
digit   significant   exact   detail accurate

```
A U P R E C I S I O N H M E
C M E A S U R E M E N T A N
C B Y G H D I G I T I O R I
U P L O K E V F E R G B T M
R B G T R T O I P L K Y I R
A C C U R A C Y A S D F A E
T Q W E R I T Y U H N I N T
E X A C T L E V E L Z X C E
Y S I G N I F I C A N T A D
```

What's an alien's favorite candy? **Martian** - mallows

10

Practice A
Perimeter and Circumference

Find the perimeter of each polygon.

1. 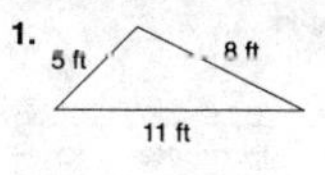

2. 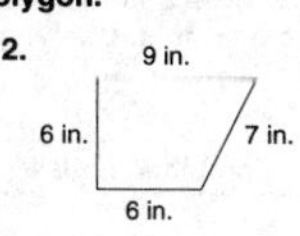

3. 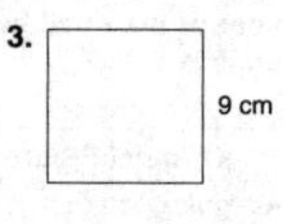

_____24 ft_____   _____28 in._____   _____36 cm_____

Find the perimeter of each rectangle.

4. 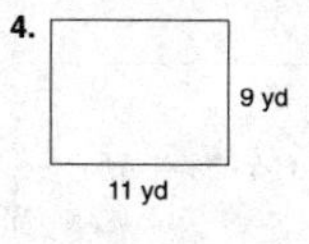

5. 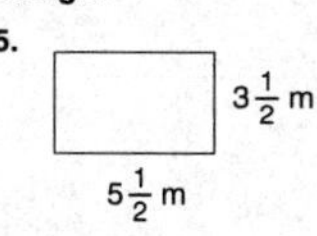

6. 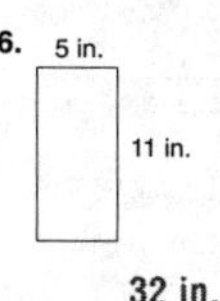

_____40 yd_____   _____18 m_____   _____32 in._____

Find the circumference of each circle to the nearest tenth. Use 3.14 for π. Choose the letter for the best answer.

7. 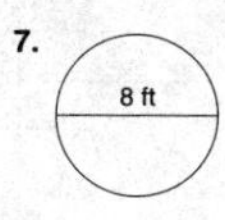

8. 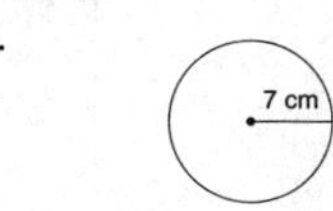

A 12.5 ft   C 25.1 ft   F 21.9 cm   H 43.9 cm
B 12.6 ft   D 25.2 ft   G 22.0 cm   J 44.0 cm

9.

10.

A 8.1 yd   C 25.2 yd   F 56.5 in.   H 28.2 in.
B 15.7 yd   D 31.4 yd   G 28.3 in.   J 14.1 in.

11. The diameter of a clock is 7 inches. What is the circumference of the clock? Use $\frac{22}{7}$ for π.

_____22 in._____

11

Practice B
Perimeter and Circumference

Find the perimeter of each polygon.

1. 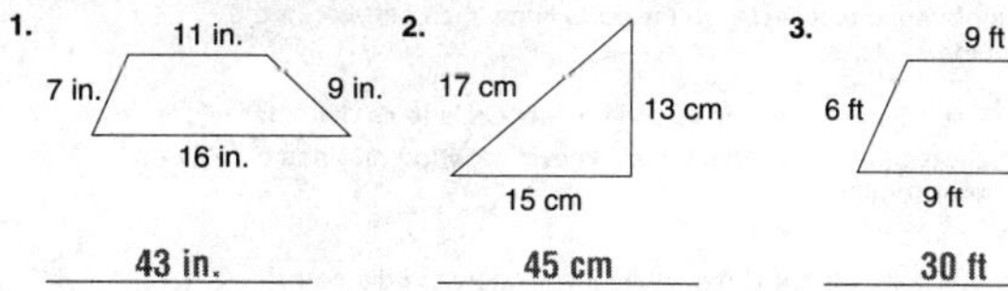

2.

3.

_____43 in._____   _____45 cm_____   _____30 ft_____

Find the perimeter of each rectangle.

4. 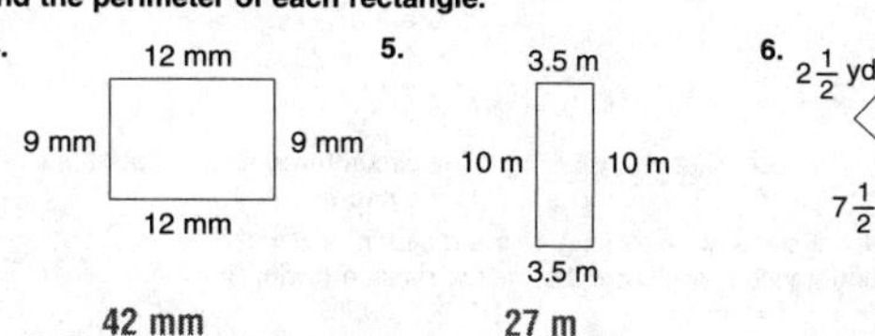

5.

6.

_____42 mm_____   _____27 m_____   _____20 yd_____

Find the circumference of each circle to the nearest tenth. Use 3.14 for π or $\frac{22}{7}$.

7. 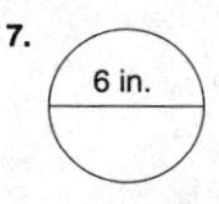

8.

9.

_____18.8 in._____   _____56.5 cm_____   _____4.7 ft_____

10. A circular swimming pool is 21 feet in diameter. What is the circumference of the swimming pool? Use $\frac{22}{7}$ for π.

_____66 ft_____

11. A jar lid has a diameter of 42 millimeters. What is the circumference of the lid? Use $\frac{22}{7}$ for π.

_____132 mm_____

12. A frying pan has a radius of 14 centimeters. What is the circumference of the frying pan? Use $\frac{22}{7}$ for π.

_____88 cm_____

12

Practice C
Perimeter and Circumference

Find the perimeter of each polygon.

1. 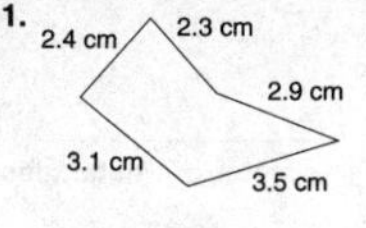

2. 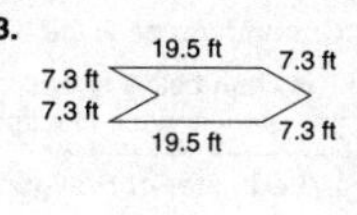

3.

_____14.2 cm_____   _____25 in._____   _____68.2 ft_____

Find the perimeter of each rectangle.

4. 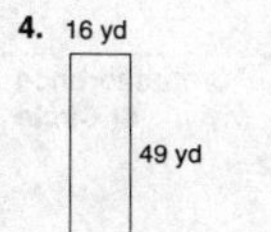

5. 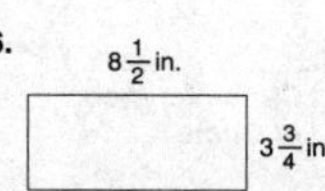

6.

_____130 yd_____   _____52.2 m_____   _____$24\frac{1}{2}$ in._____

Find each missing measurement to the nearest tenth. Use 3.14 for π or $\frac{22}{7}$.

7. $r = 14.5$ in.

$d = $ _____29 in._____

$C = $ _____91.1 in._____

8. $r = $ _____$5\frac{2}{3}$ yd_____

$d = $ _____$11\frac{1}{3}$ yd_____

$C = 35.6$ yd

9. $r = $ _____3.8 cm_____

$d = 7.6$ cm

$C = $ _____23.9 cm_____

10. $r = $ _____1.8 m_____

$d = $ _____3.6 m_____

$C = 11.3$ m

Solve.

11. In NCAA basketball rules, the basketball can have a maximum circumference of 30 inches. What is the maximum diameter of an NCAA basketball to the nearest hundredth?

_____9.55 in._____

13

Reteach
Perimeter and Circumference

The **perimeter** of any figure is the distance around the figure. Think of *perimeter* as "going around the rim" of a figure. Find the perimeter of a figure by adding the measures of the sides.

Find the perimeter of each polygon.

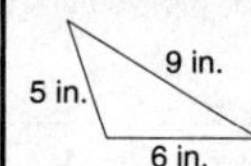

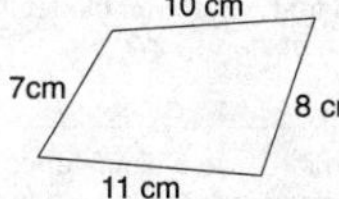

$P = 6 + 5 + 9$
$P = 20$
The perimeter of the triangle is 20 in.

$P = 7 + 11 + 10 + 8$
$P = 36$
The perimeter of the quadrilateral is 36 cm.

To find the perimeter of a rectangle, you can add the length and the width and multiply the sum by 2. The formula for the perimeter of a rectangle is $P = 2(\ell + w)$.

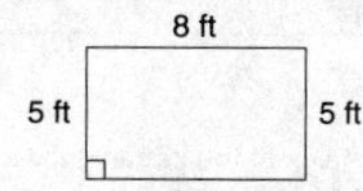

$P = 2(\ell + w)$
$P = 2(8 + 5)$
$P = 2(13)$
$P = 26$
The perimeter of the rectangle is 26 feet.

Find the perimeter of each polygon.

1. 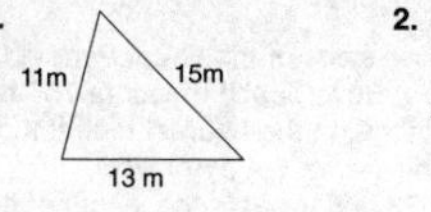

2.

3.

$P = $ __11__ + __13__ + __15__
$P = $ __39__ m

$P = $ __21__ + __23__
     + __24__ + __30__
$P = $ __98__ ft

$P = 2($ __16__ + __8__ $)$
$P = $ __48__ cm

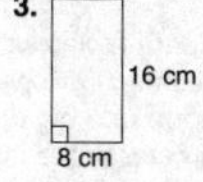

14

Reteach
Perimeter and Circumference (continued)

The distance around a circle is called the **circumference**. To find the circumference of a circle, you need to know the diameter or the radius of the circle.

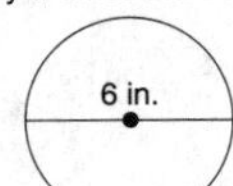

The ratio of the circumference of any circle to its diameter $\left(\frac{C}{d}\right)$ is always the same. This ratio is known as π (pi) and has a value of approximately 3.14.

To find the circumference C of a circle if you know the diameter d, multiply π times the diameter. $C = \pi \cdot d$, or $C \approx 3.14 \cdot d$.

6 in.

$C = \pi \cdot d$
$C \approx 3.14 \cdot d$
$C \approx 3.14 \cdot 6$
$C \approx 18.84$
The circumference is about 18.8 in. to the nearest tenth.

The diameter of a circle is twice as long as the radius r, or $d = 2r$.
To find the circumference if you know the radius, replace d with $2r$ in the formula. $C = \pi \cdot d = \pi \cdot 2r$

Find the circumference given the diameter.

4. $d = 9$ cm
$C = \pi \cdot d$
$C \approx 3.14 \cdot \underline{\quad 9 \quad}$
$C \approx \underline{\quad 28.26 \quad}$
The circumference is $\underline{\quad 28.3 \quad}$ cm to the nearest tenth of a centimeter.

Find the circumference given the radius.

5. $r = 13$ in.
$C = \pi \cdot 2r$
$C \approx 3.14 \cdot (2 \cdot \underline{\quad 13 \quad})$
$C \approx 3.14 \cdot \underline{\quad 26 \quad}$
$C \approx \underline{\quad 81.64 \quad}$
The circumference is $\underline{\quad 81.6 \quad}$ in. to the nearest tenth of an inch.

Find the circumference of each circle to the nearest tenth. Use 3.14 for π.

6.
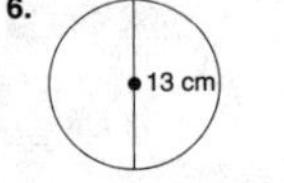
13 cm

$\underline{\quad 40.8 \text{ cm} \quad}$

7.
5 ft

$\underline{\quad 31.4 \text{ ft} \quad}$

8.
1.5 in.

$\underline{\quad 9.4 \text{ in.} \quad}$

Holt Mathematics

Challenge
All-Around Formulas

Use the first figure in each row to write a formula for the perimeter of the combined figure next to it. Use your formulas in Exercises 5–8.

	Original Figure	Combined Figure	Perimeter
1.	A regular octagon: m		$P = 26m$
2.	An isosceles triangle: s, b		$P = 2s + 6b$
3.	A parallelogram: w, ℓ		$P = 4\ell + 4w$
4.	A semicircle: d		$P = 3\left(\frac{1}{2}\pi d + d\right)$

5. What is the perimeter of the combined figure in Exercise 1 if $m = 4$ in.? $\underline{\quad 104 \text{ in.} \quad}$

6. What is the perimeter of the combined figure in Exercise 2 if $s = 4.5$ m and $b = 5.2$ m? $\underline{\quad 40.2 \text{ m} \quad}$

7. What is the length of the original figure in Exercise 3 if the width is 6 in. and the perimeter of the combined figure is 56 in.? $\underline{\quad 8 \text{ in.} \quad}$

8. What is the perimeter of the combined figure in Exercise 4 if $d = 10$ cm? $\underline{\quad 77.1 \text{ cm} \quad}$

Holt Mathematics

Problem Solving
Perimeter and Circumference

Write the correct answer.

1. Mr. Marcos, the gym teacher, had the seventh graders run around the perimeter of the gym 3 times. The gym has a length of 34 feet and a width of 58 feet. What was the total distance the students ran?

$\underline{\quad 552 \text{ ft} \quad}$

2. The distance between bases on a baseball field is 90 feet. If 3 players hit home runs during a game and each runs around all 4 bases, what is the total distance the players run?

$\underline{\quad 1{,}080 \text{ ft} \quad}$

3. Basketball rims have a diameter of 18 inches. If you want to put a band around a basketball rim, how much material to the nearest tenth of an inch will you need?

$\underline{\quad 56.5 \text{ in.} \quad}$

4. A pizza cutter has a diameter of 2.5 inches. To cut a pizza in half, the cutter makes two complete revolutions. What is the diameter of the pizza?

$\underline{\quad 15.7 \text{ in.} \quad}$

5. A round stained-glass window has a circumference of 195 inches. What is the radius of the window to the nearest inch?

$\underline{\quad 31 \text{ in.} \quad}$

6. A planter full of pansies has a diameter of 14 inches. What is the circumference of the planter to the nearest inch?

$\underline{\quad 44 \text{ in.} \quad}$

Choose the letter of the correct answer.

7. A welcome mat on the front porch is a semicircle. The straight side of the mat is 36 inches. What is the perimeter of the mat?
A 92.52 in.   C 56.52 in.
B 64.26 in.   D 28.26 in.

8. The radius of the planet Jupiter is about 44,368 miles. What is the approximate circumference of Jupiter to the nearest mile?
F 557,262 mi   H 139,316 mi
G 278,631 mi   J 69,658 mi

9. Four square tables with sides of 48 inches each are placed end to end to form one big table. What is the perimeter of the table that is formed?
A 192 in.   C 480 in.
B 384 in.   D 768 in.

10. Three sides of the Great Pyramid at Giza, Egypt, each measure 756 feet in length to the nearest foot. If the perimeter of the pyramid is 3,023 feet, what is the length of the fourth side of the pyramid?
F 754 ft   H 756 ft
G 755 ft   J 757 ft

Holt Mathematics

Reading Strategies
Use A Graphic Organizer

Perimeter is the distance around a polygon.

Circumference is the distance around a circle.

The chart below shows formulas for finding the perimeter of polygons and the circumference of circles.

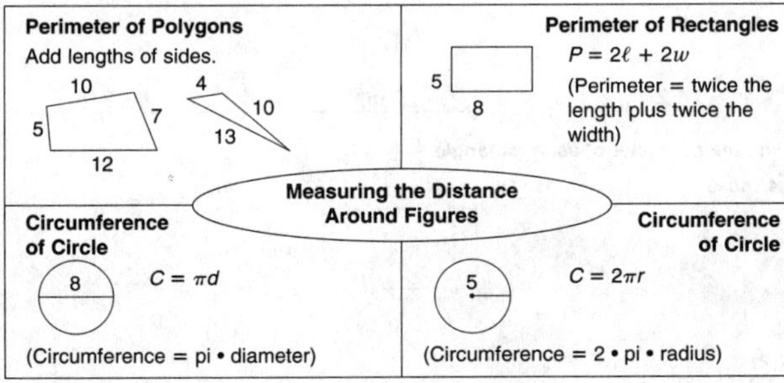

Use the information in the chart above to complete each exercise.

1. Write the lengths for the triangle shown above.

$\underline{\quad 4, 13, \text{ and } 10 \quad}$

2. How would you find the perimeter of the triangle?

$\underline{\quad \text{Add the lengths of the three sides.} \quad}$

3. If you knew the radius of a circle, what formula would you use to find its circumference?

$\underline{\quad C = 2\pi r \quad}$

4. Write the formula for finding the perimeter of a rectangle.

$\underline{\quad 2\ell + 2w \quad}$

5. If you knew the diameter of a circle, what formula would you use to find its circumference?

$\underline{\quad C = \pi d \quad}$

Holt Mathematics

Holt Mathematics

Puzzles, Twisters & Teasers
Around and Around!

Find the perimeter or circumference of each figure. Round the circumference to the nearest tenth. Match your answers to the corresponding letter to solve the riddle.

1. __44 m__ = E

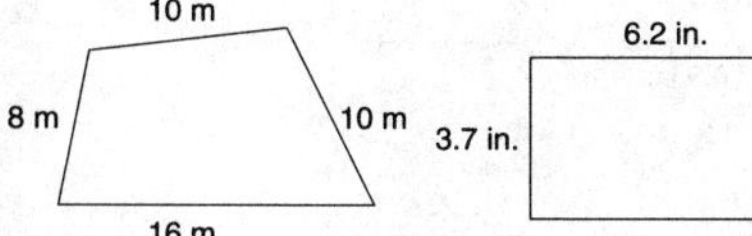

2. __19.8 in.__ = V

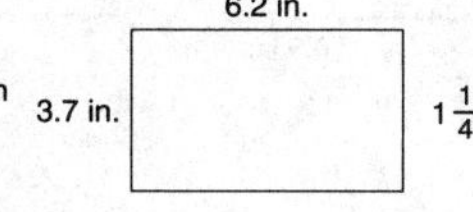

3. __10 ft__ = L

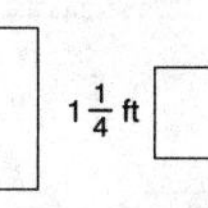

4. __25.1 cm__ = I

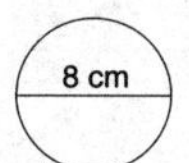

5. __19.5 cm__ = C

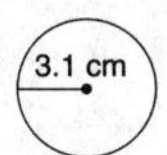

6. __18 m__ = O

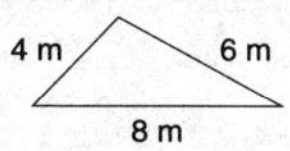

7. __24 in.__ = R

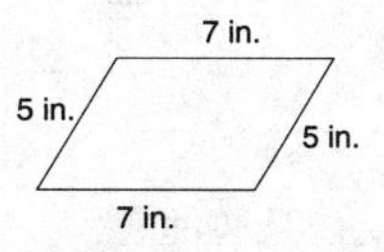

8. __30 ft__ = S

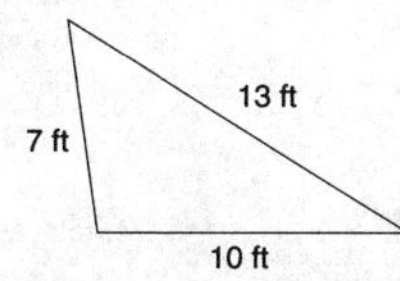

9. __9.4 ft__ = U

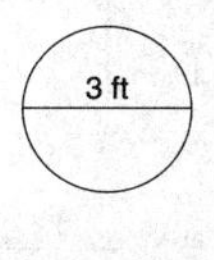

What's round and bad-tempered?

A	V	I	C	I	O	U	S
	19.8	25.1	19.5	25.1	18	9.4	30

C	I	R	C	L	E
19.5	25.1	24	19.5	10	44

19
Holt Mathematics

Practice A
Area of Parallelograms

Find the area of each figure. Choose the letter for the best answer.

1.

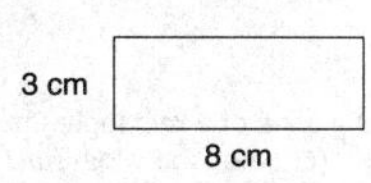
3 cm / 8 cm

2.

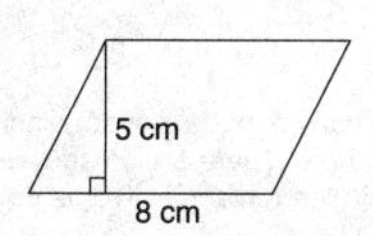
5 cm / 8 cm

A 11 cm²   Ⓒ 24 cm²   F 26 cm²   H 48 cm²
B 22 cm²   D 48 cm²   Ⓖ 40 cm²   J 80 cm²

Find the area of each rectangle.

3.

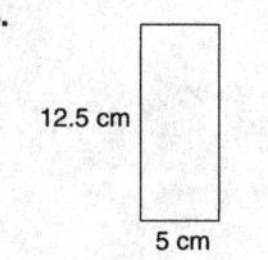
12.5 cm / 5 cm

4.

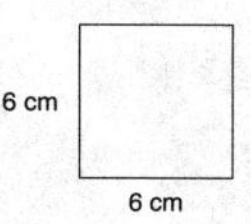
6 cm / 6 cm

5.

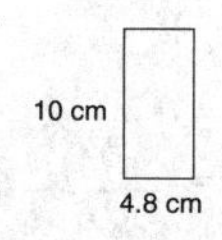
10 cm / 4.8 cm

__62.5 cm²__   __36 cm²__   __48 cm²__

Find the area of each parallelogram.

6.

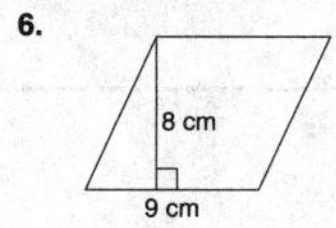
8 cm / 9 cm

7.

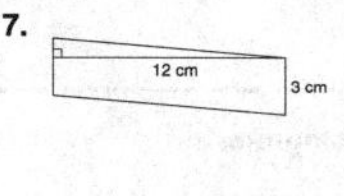
12 cm / 3 cm

8.

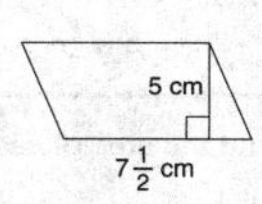
5 cm / 7½ cm

__72 cm²__   __36 cm²__   __$37\frac{1}{2}$ cm²__

9. Michelle wants to carpet her living room. The area of the living room is 192 ft². The length of the living room is 16 ft. What is the width of the living room? __12ft__

10. Mustafa is tiling his bathroom. The section that needs to be tiled is 62 in. by 70 in. How many square inches of tile does he need? __4,340 in²__

20
Holt Mathematics

Practice B
Area of Parallelograms

Find the area of each rectangle.

1.

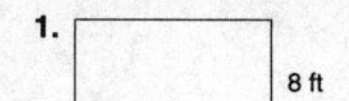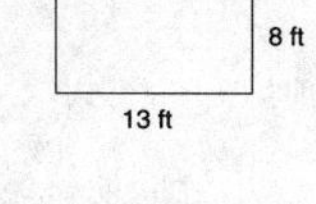
8 ft / 13 ft

2.

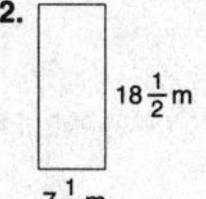
18½ m / 7½ m

3.

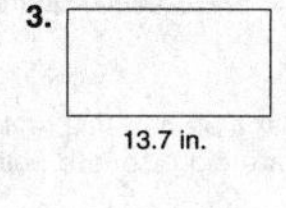
7.3 in. / 13.7 in.

__104 ft²__   __$138\frac{3}{4}$ m²__   __100.01 in²__

Find the area of each parallelogram.

4.

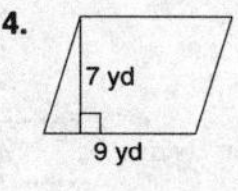
7 yd / 9 yd

5.

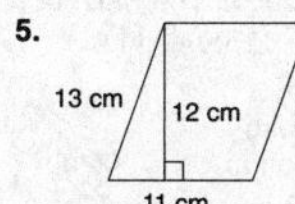
13 cm / 12 cm / 11 cm

6.

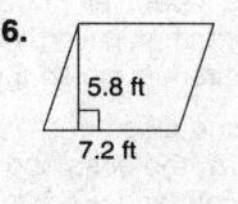
5.8 ft / 7.2 ft

__63 yd²__   __132 cm²__   __41.76 ft²__

7.

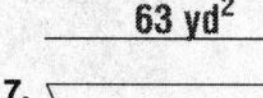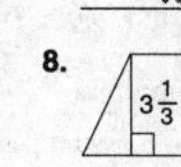
2.5 m / 6 m

8.

3⅓ in. / 10½ in.

9.

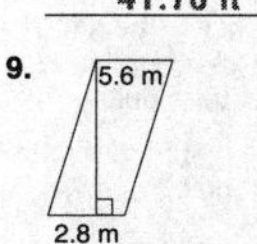
5.6 m / 2.8 m

__15 m²__   __35 in.²__   __15.68 m²__

10. A dollar bill is 15.5 cm long and 6.5 cm wide. What is the area of a dollar bill?

__100.75 cm²__

11. A rectangular hallway has an area of 70 ft2. The width of the hallway is 4 feet. What is the length of the hallway?

__17.5 ft__

21
Holt Mathematics

Practice C
Area of Parallelograms

Find the area of each rectangle.

1.

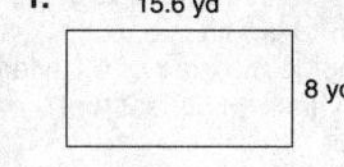
15.6 yd / 8 yd

2.

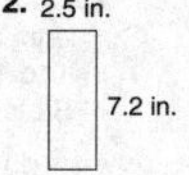
2.5 in. / 7.2 in.

3.

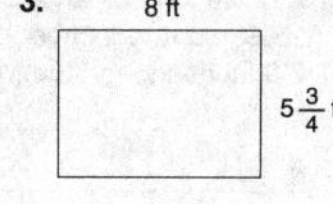
8 ft / $5\frac{3}{4}$ ft

__124.8 yd²__   __18 in.²__   __46 ft²__

Find the area of each parallelogram.

4.

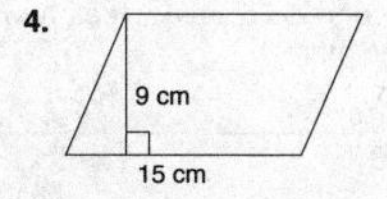
9 cm / 15 cm

5.

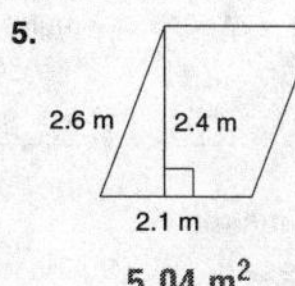
2.6 m / 2.4 m / 2.1 m

6.

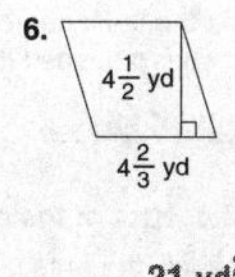
4½ yd / 4⅔ yd

__135 cm²__   __5.04 m²__   __21 yd²__

Find the area of each polygon.

7. rectangle: ℓ = 29 in., w = 15 in.

__435 in²__

8. parallelogram: b = 8 m, h = 3.6 m

__28.8 m²__

Sketch the figure with the given vertices. Then find the area of the figure.

9. (3, −2), (8, −2), (3, 6), (8, 6)

__40 units²__

10. (2, 0), (5, 3), (8, 0), (11, 3)

__18 units²__

11. (−4,1), (−9, 1), (0, 7), (−5, 7)

__30 units²__

12. (−1, −2), (4, −2), (−1, 5), (4, 5)

__35 units²__

13. What is the length of the base of a parallelogram with an area of 36 cm² and a height of 4.5 cm?

__8 cm__

14. A window is composed of 6 windowpanes. Each has a length of 10½ in. and a width of 7¼ in. What is the total area of glass in the window?

__$456\frac{3}{4}$ in²__

22
Holt Mathematics

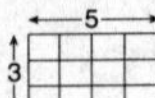 **Reteach**
9-3 Area of Parallelograms

The **area** of a figure is the number of square units inside the figure.

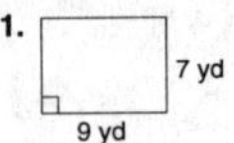

You can count the squares inside the rectangle. There are 15 square units within the rectangle. This is equal to 5 • 3.

To find the area of a rectangle, multiply the length (ℓ) times the width (w).
$$A = \ell \cdot w$$

Find the area of each rectangle.

1. 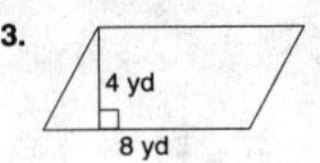

$A = l \cdot w$

$A = \underline{9} \cdot \underline{7}$

$A = \underline{63}$

The area is $\underline{63}$ yd^2.

2. 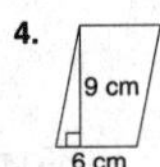

$A = l \cdot w$

$A = \underline{12} \cdot \underline{5}$

$A = \underline{60}$

The area is $\underline{60}$ in^2.

To find the area of a parallelogram, multiply the base b times the height h.
$$A = b \cdot h$$

Find the area of each parallelogram.

3.

$A = b \cdot h$

$A = \underline{8} \cdot \underline{4}$

$A = \underline{32}$

The area is $\underline{32}$ yd^2.

4.

$A = b \cdot h$

$A = \underline{6} \cdot \underline{9}$

$A = \underline{54}$

The area is $\underline{54}$ cm^2.

 23 **Holt Mathematics**

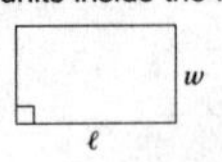 **Challenge**
9-3 Size It Up!

Gonzalez Builders purchased the four lots shown below. The company intends to build homes on the lots and needs to know the total area in square feet of each lot. They also need to know the perimeter in feet of each lot.

You can use proportions to calculate the areas and perimeters. For each figure, the scale is 1 inch: 60 feet.

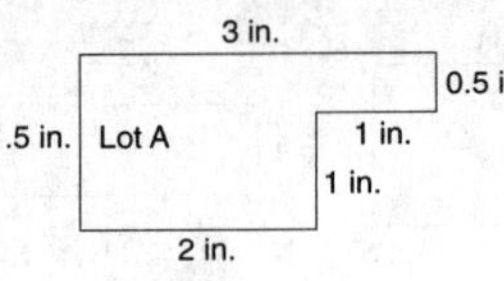

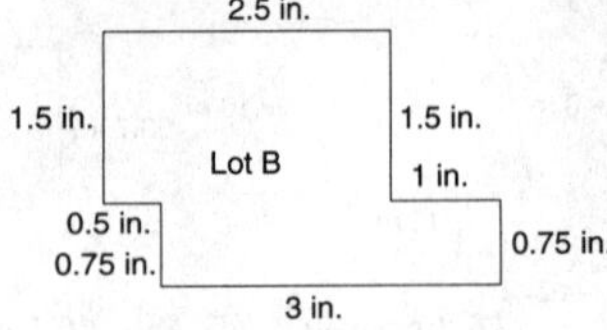

1. total area $\underline{12{,}600 \text{ ft}^2}$

   perimeter $\underline{540 \text{ ft}}$

2. total area $\underline{21{,}600 \text{ ft}^2}$

   perimeter $\underline{690 \text{ ft}}$

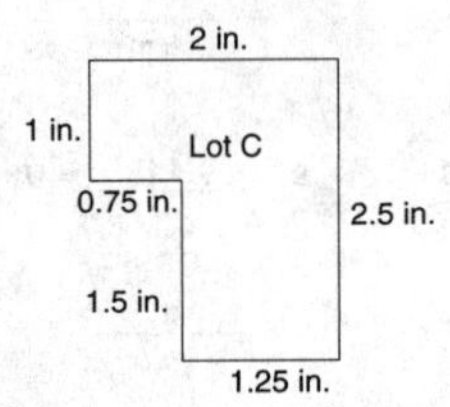

3. total area $\underline{13{,}950 \text{ ft}^2}$

   perimeter $\underline{540 \text{ ft}}$

4. total area $\underline{23{,}400 \text{ ft}^2}$

   perimeter $\underline{690 \text{ ft}}$

5. What is the combined area of all four lots? $\underline{71{,}550 \text{ ft}^2}$

6. What length of fencing would be needed to enclose all four lots during construction? $\underline{2{,}460 \text{ ft}}$

 24 **Holt Mathematics**

 Problem Solving
9-3 Area of Parallelograms

Write the correct answer.

1. A dollar bill has an area of 15.86 square inches. If a dollar bill is 2.6 inches long, how wide is it?

_6.1 in._

2. On an official United States flag, the ratio of width to length is exactly 1 to 1.9. What is the area of a United States flag whose width is 2 feet?

_7.6 ft^2_

3. A back yard is shaped like a parallelogram with a height of 32 feet and a base of 100 feet. One bag of grass seed covers 125 square feet. What is the least number of bags of seed needed to seed the lawn?

26 bags of seed

4. The art club is painting a mural on a school wall. The mural is in the shape of a parallelogram. If the base of the mural is 10.5 feet long and the mural covers 89.25 square feet, how high is the mural?

8.5 ft high

Choose the letter of the correct answer.

5. In baseball, the area of each base is 225 square inches. Each base is a square. What is the length of each side of a base on a baseball field?

A 12 in.   C 25 in.

B 22.5 in.   (D) 15 in.

6. The area of a parallelogram is 632.1 square centimeters. Its base is 24.5 centimeters. What is the height of the parallelogram?

(F) 25.8 cm   H 21.9 cm

G 705.6 cm   J 11.8 cm

7. The official rules for volleyball were developed in 1897. The rules state that the court or floor space must be 25 feet wide and 50 feet long. An official basketball court is 94 feet by 50 feet. How much larger is the area of a basketball court than the area of a volleyball court?

A 69 ft^2 larger

(B) 3,450 ft^2 larger

C 1,250 ft^2 larger

D 4,700 ft^2 larger

8. Two parallelograms each have an area of 288 square inches. One has a height of 12 inches, and the other has a height of 18 inches. What are the bases of each parallelogram?

F 40 in. and 30 in.

G 22 in. and 15 in.

(H) 24 in. and 16 in.

J 26 in. and 20 in.

 25 **Holt Mathematics**

 Reading Strategies
9-3 Compare and Contrast

Area measures the surface a figure covers. Area is measured in square units.

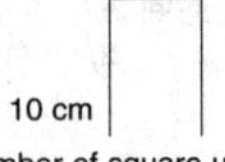

To find the area of this rectangle, count the number of square units. There are eight square units.

Another way to find the area is to use this formula: $A = \ell w$.

Read: "Area equals length times width."

4 units times 2 units = 8 square units.
Area = 8 square units

Compare a parallelogram to a rectangle. The width of a parallelogram is called the height. The length of a parallelogram is called the base.

Moving one piece of the parallelogram shows how the base and height compare to the length and width of a rectangle.

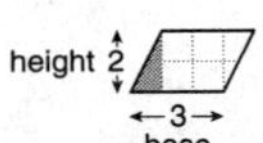

Area = base • height ($A = bh$)   Area = length • width ($A = \ell w$)
$A = 3 \cdot 2$ $A = 3 \cdot 2$
$A = 6$ square units $A = 6$ square units

1. The height of a parallelogram is the same as what part of a rectangle?

the width

2. The base of a parallelogram is the same as what part of a rectangle?

the length

3. What is the same about finding the area of a parallelogram and a rectangle?

Possible answer: You multiply either the length or base times
the height or width.

4. What is different about finding the area of a parallelogram and a rectangle?

A rectangle has length and width; a parallelogram has base and height.

 26 **Holt Mathematics**

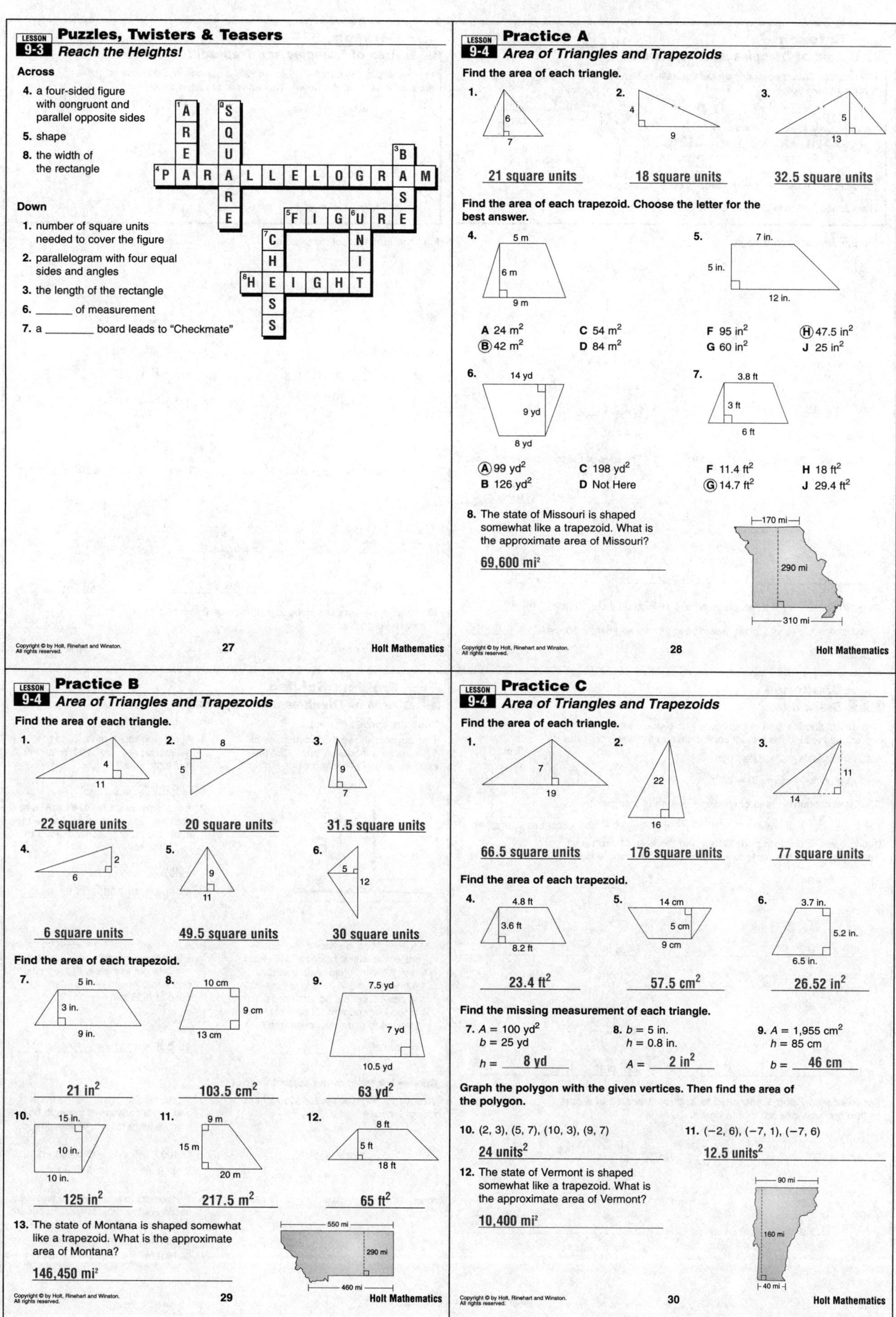

LESSON 9-3
Puzzles, Twisters & Teasers
Reach the Heights!

Across
4. a four-sided figure with oongruont and parallel opposite sides
5. shape
8. the width of the rectangle

Down
1. number of square units needed to cover the figure
2. parallelogram with four equal sides and angles
3. the length of the rectangle
6. ______ of measurement
7. a ________ board leads to "Checkmate"

A R E A
S Q U A R E
P A R A L L E L O G R A M
B A S E
F I G U R E
C H E S S
U N I T
H E I G H T

LESSON 9-4
Practice A
Area of Triangles and Trapezoids

Find the area of each triangle.

1. 6, 7 — 21 square units
2. 4, 9 — 18 square units
3. 5, 13 — 32.5 square units

Find the area of each trapezoid. Choose the letter for the best answer.

4. 5 m, 6 m, 9 m
A 24 m² C 54 m²
B 42 m² D 84 m²

5. 7 in., 5 in., 12 in.
F 95 in² H 47.5 in²
G 60 in² J 25 in²

6. 14 yd, 9 yd, 8 yd
A 99 yd² C 198 yd²
B 126 yd² D Not Here

7. 3.8 ft, 3 ft, 6 ft
F 11.4 ft² H 18 ft²
G 14.7 ft² J 29.4 ft²

8. The state of Missouri is shaped somewhat like a trapezoid. What is the approximate area of Missouri?
69,600 mi²
170 mi, 290 mi, 310 mi

LESSON 9-4
Practice B
Area of Triangles and Trapezoids

Find the area of each triangle.

1. 4, 11 — 22 square units
2. 8, 5 — 20 square units
3. 9, 7 — 31.5 square units
4. 6, 2 — 6 square units
5. 9, 11 — 49.5 square units
6. 5, 12 — 30 square units

Find the area of each trapezoid.

7. 5 in., 3 in., 9 in. — 21 in²
8. 10 cm, 9 cm, 13 cm — 103.5 cm²
9. 7.5 yd, 7 yd, 10.5 yd — 63 yd²
10. 15 in., 10 in., 10 in. — 125 in²
11. 9 m, 15 m, 20 m — 217.5 m²
12. 8 ft, 5 ft, 18 ft — 65 ft²

13. The state of Montana is shaped somewhat like a trapezoid. What is the approximate area of Montana?
146,450 mi²
550 mi, 290 mi, 460 mi

LESSON 9-4
Practice C
Area of Triangles and Trapezoids

Find the area of each triangle.

1. 7, 19 — 66.5 square units
2. 22, 16 — 176 square units
3. 11, 14 — 77 square units

Find the area of each trapezoid.

4. 4.8 ft, 3.6 ft, 8.2 ft — 23.4 ft²
5. 14 cm, 5 cm, 9 cm — 57.5 cm²
6. 3.7 in., 5.2 in., 6.5 in. — 26.52 in²

Find the missing measurement of each triangle.

7. A = 100 yd²
b = 25 yd
h = 8 yd

8. b = 5 in.
h = 0.8 in.
A = 2 in²

9. A = 1,955 cm²
h = 85 cm
b = 46 cm

Graph the polygon with the given vertices. Then find the area of the polygon.

10. (2, 3), (5, 7), (10, 3), (9, 7)
24 units²

11. (−2, 6), (−7, 1), (−7, 6)
12.5 units²

12. The state of Vermont is shaped somewhat like a trapezoid. What is the approximate area of Vermont?
10,400 mi²
90 mi, 160 mi, 40 mi

Reteach
Area of Triangles and Trapezoids

The diagram shows how you can cut a parallelogram into two congruent triangles.

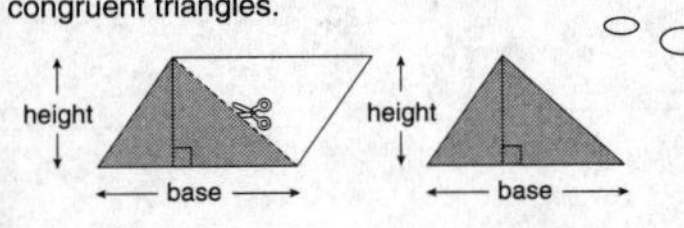

The area of the triangle is $\frac{1}{2}$ the area of the parallelogram.

The formula for the area of a triangle is $A = \frac{1}{2} \cdot b \cdot h$.

Find the area of each triangle.

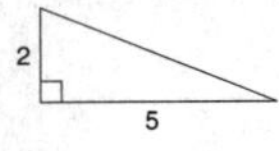

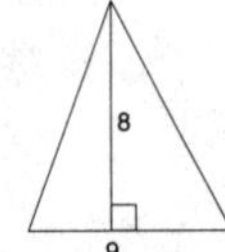

1. $A = \frac{1}{2} \cdot b \cdot h$

   $A = \frac{1}{2} \cdot \underline{5} \cdot \underline{2}$

   $A = \frac{1}{2} \cdot \underline{10}$

   $A = \underline{5}$

   The area of the triangle is $\underline{5}$ units^2.

2. $A = \frac{1}{2} \cdot b \cdot h$

   $A = \frac{1}{2} \cdot \underline{9} \cdot \underline{8}$

   $A = \frac{1}{2} \cdot \underline{72}$

   $A = \underline{36}$

   The area of the triangle is $\underline{36}$ units^2.

3. 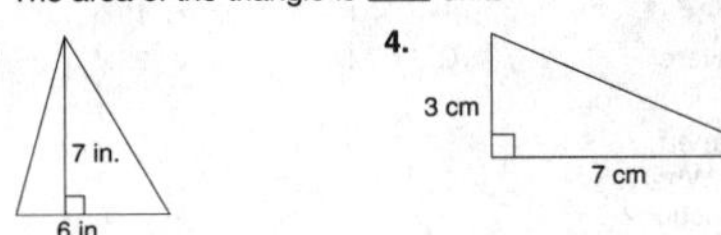

3. ______ 21 in^2

4. ______ 10.5 cm^2

5. ______ 18 yd^2

6. What is the area of a triangle with base 16 m and height 10 m? __ 80 m^2

7. What is the area of a triangle with base 25 mm and height 50 mm? __ 625 mm^2

Holt Mathematics

Reteach
Area of Triangles and Trapezoids (continued)

In a trapezoid, the parallel sides are called the *bases*. One base is always longer than the other. The bases are labeled base 1 and base 2.

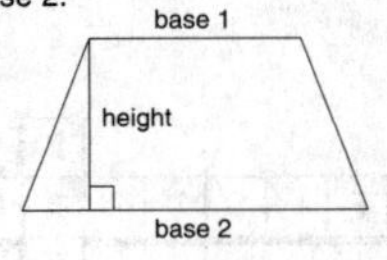

Area of trapezoid = $\frac{1}{2}h(b_1 + b_2)$

Find the area of each trapezoid.

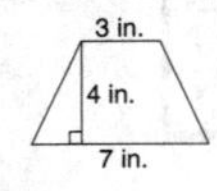

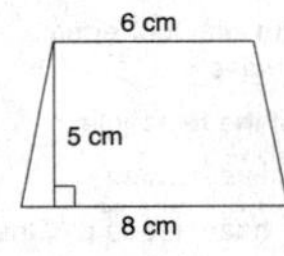

8. $A = \frac{1}{2}h(b_1 + b_2)$

   $A = \frac{1}{2} \cdot \underline{4} \, (\, \underline{3} \, + \, \underline{7} \,)$

   $A = \frac{1}{2} \cdot \underline{4} \, (\, \underline{10} \,)$

   $A = \frac{1}{2} \cdot \underline{40}$

   $A = \underline{20}$

   The area of the trapezoid is $\underline{20}$ in^2.

9. $A = \frac{1}{2}h(b_1 + b_2)$

   $A = \frac{1}{2} \cdot \underline{5} \, (\, \underline{6} \, + \, \underline{8} \,)$

   $A = \frac{1}{2} \cdot \underline{5} \, (\, \underline{14} \,)$

   $A = \frac{1}{2} \cdot \underline{70}$

   $A = \underline{35}$

   The area of the trapezoid is $\underline{35}$ cm^2.

10. 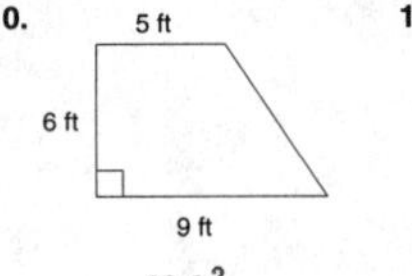

11. 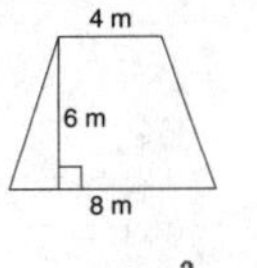

12. 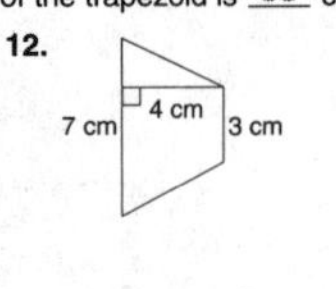

10. ______ 42 ft^2

11. ______ 36 m^2

12. ______ 20 cm^2

13. What is the area of a trapezoid with bases 25 yd and 75 yd and height 10 yd? ______ 500 yd^2

Holt Mathematics

Challenge
Break It Up

One way to find the area of this figure is to divide it into a rectangle and a triangle. Find the area of each, then add the areas together.

Area of rectangle = $7 \cdot 10 = 70$ m^2

Area of triangle = $\frac{1}{2}(5 \cdot 10) = 25$ m^2

Total area of figure = Area of rectangle + Area of triangle

= 70 m^2 + 25 m^2 = 95 m^2

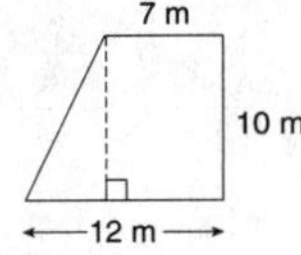

Divide each figure into parts. Then find the area of each part. Add the areas of the parts to find the area of the whole figure.

1. 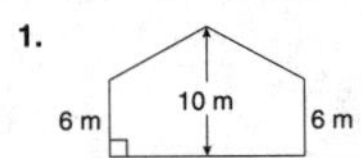

   ______ 120 m^2

2. 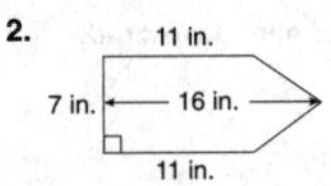

   ______ 94.5 in^2

3. 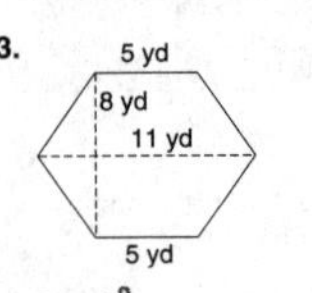

   ______ 64 yd^2

4. 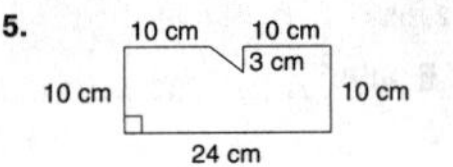

   ______ 224 cm^2

For Exercises 5 and 6, you need to *subtract* the area of a part to find the area of each whole figure.

5. 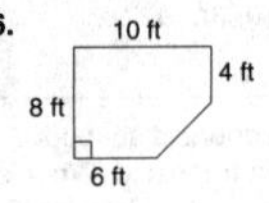

   ______ 234 cm^2

6.

   ______ 72 ft^2

Holt Mathematics

Problem Solving
Area of Triangles and Trapezoids

Write the correct answer.

The diagram shows the dimensions of the sails on a model sailboat. Use the diagram to solve Problems 1–2.

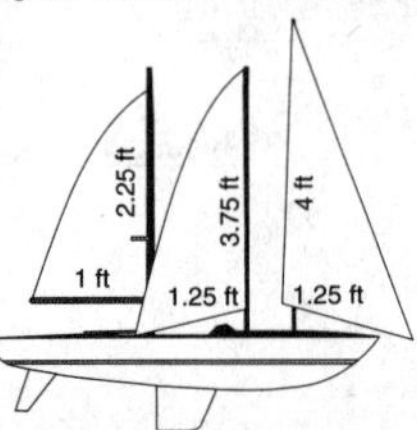

1. About how much material to the nearest square foot will be needed to make the sails?

   __ about 6 ft^2

2. If the dimensions for each sail were doubled, how would that change the total amount of material needed to make the sails?

   __ You would need 4 times as much, or about 24 ft^2.

3. A flower bed is shaped like a trapezoid with a height of 3.5 yards, one 2.8-yard base, and another 4.6-yard base. A packet of flower seeds covers 5.6 square yards. What is the least number of packets needed to plant the flower bed?

   __ 3 packets

4. A triangular road sign has a height of 8 feet and a base of 16.5 feet. How much larger in area is this sign than one with a height of 4 feet and a base of 8.25 feet?

   __ 49.5 ft^2 larger

Choose the letter of the correct answer.

This diagram shows the top view of the roof of a house.

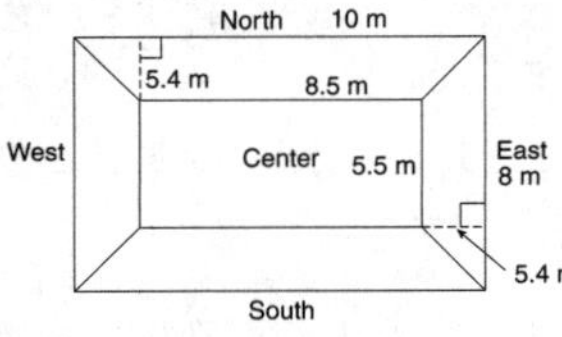

5. If you need to reshingle the north and south sections of the roof, how many square meters of shingles will you need?

   A 199.8 m^2 C 49.95 m^2
   Ⓑ 99.9 m^2 D 459 m^2

6. If you need to reshingle the west section of the roof, how many square meters of shingles will you need?

   F 13.5 m^2 Ⓗ 36.45 m^2
   G 18.9 m^2 J 72.9 m^2

Holt Mathematics

Holt Mathematics

Reading Strategies
Use Graphic Aid

Knowing how to find the area of a square or parallelogram can help you find the area of a triangle.

When you draw a diagonal line through a square, you make two equal triangles.

Answer each question.

1. How many square units are in the square? _9_

2. To find out how many squares are in one of the triangles, count the number of half squares first. How many did you find? _3_

3. How many whole squares can you count? _3_

4. How many half and whole squares together? $4\frac{1}{2}$

You can also draw a diagonal line through a parallelogram and divide it into two equal triangles.

Use the figure to answer each.

5. How many square units are in the parallelogram? _8_

6. Count the full squares and half squares in one triangle and write the total. _4_

7. Compare the area of a parallelogram or square with the area of a triangle.

   The area of a triangle is half the area of a parallelogram or square.

35
Holt Mathematics

Puzzles, Twisters & Teasers
All Washed Up!

Find the area of each figure below. Match the letters to solve the riddle.

1. __28__ S
2. __6__ M
3. __15__ F

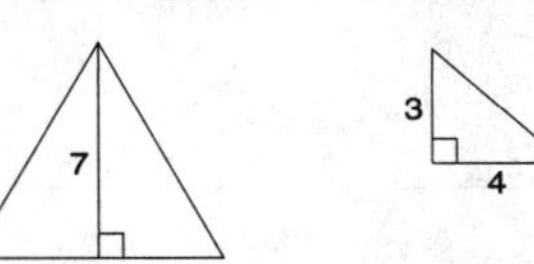
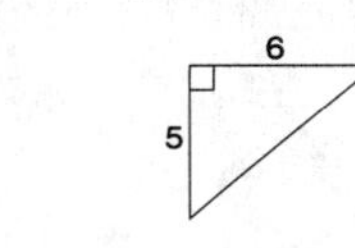

4. __32 in__ I
5. __187 cm__ V
6. __25__ C

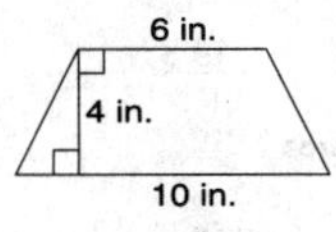
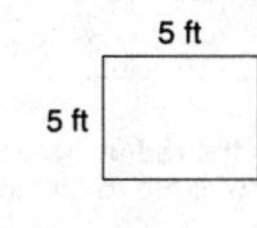

7. __64 m__ A
8. __20 m__ R
9. __54 ft__ W

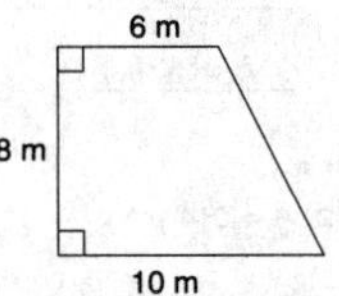
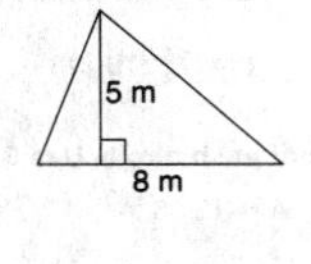
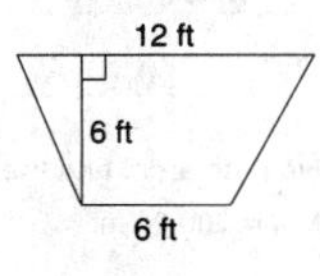

What washes up on small beaches?

M	I	C	R	o	-	W	A	V	E	S
6	32	25	20			54	64	187	15	28

36
Holt Mathematics

Practice A
Area of Circles

Find the area of each circle to the nearest tenth. Use 3.14 for π. Cross out each number in the box that matches an area.

19.6	254.3	50.2	78.5	22.0	260.0
153.9	28.3	3.1	379.9	171.9	72.3	176.6

1. 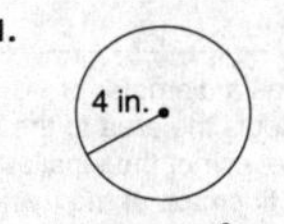
   50.2 in²

2. 153.9 m²

3. 254.3 yd²

4. 3.1 cm²

5. 171.9 km²

6. 78.5 in²

7. 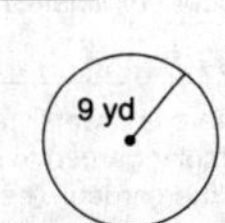
   379.9 mm²

8. 19.6 ft²

9. 28.3 m²

10. A round table has a diameter of 28 inches. What is the area of the tabletop? Use $\frac{22}{7}$ for π. 616 in²

11. Find the area of the shaded region of the circle. Use 3.14 for π. Round your answer to the nearest tenth. 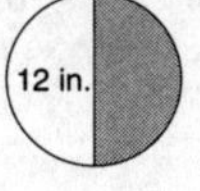

    113.0 in²

37
Holt Mathematics

Practice B
Area of Circles

Find the area of each circle to the nearest tenth. Use 3.14 for π.

1. 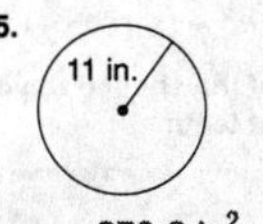
   113.0 m²

2. 50.2 ft²

3. 153.9 yd²

4. 19.6 cm²

5. 379.9 in²

6. 28.3 mm²

7. 78.5 in²

8. 132.7 cm²

9. 162.8 yd²

10. A Susan B. Anthony dollar coin has a diameter of 26.50 millimeters. What is the area of the coin to the nearest hundredth? 551.27 mm²

11. A tablecloth for a round table has a radius of 21 inches. What is the area of the tablecloth? Use $\frac{22}{7}$ for π. 1,386 in²

12. Use a centimeter ruler to measure the radius of the circle. Then find the area of the shaded region of the circle. Use 3.14 for π. Round your answer to the nearest tenth. 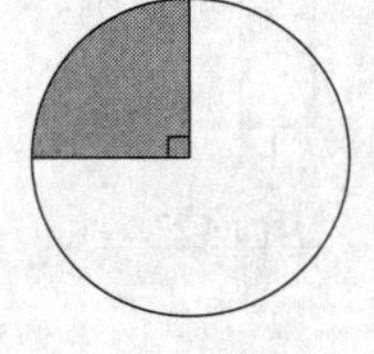

    3.8 cm²

38
Holt Mathematics

Find the area of each circle to the nearest tenth. Use 3.14 for π.

1. 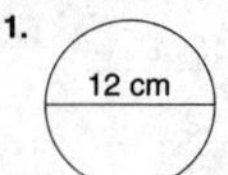
12 cm

113.0 cm²

2.
9.5 in.

283.4 in²

3.
7 yd

38.5 yd²

4.
1.4 in.

6.2 in²

5.
7.6 cm

45.3 cm²

6.
2.5 ft

19.6 ft²

Given the radius or diameter, find the circumference and area of each circle to the nearest tenth. Use 3.14 for π.

7. $r = 4$ in.

$C = 25.1$ in.

$A = 50.2$ in²

8. $d = 14$ cm

$C = 44.0$ cm

$A = 153.9$ cm²

9. $d = 26$ ft

$C = 81.6$ ft

$A = 530.7$ ft²

Given the area, find the radius of each circle. Use 3.14 for π.

10. $A = 200.96$ m²

$r = 8$ m

11. $A = 12.56$ in²

$r = 2$ in.

12. $A = 314$ yd²

$r = 10$ yd

13. A playground area is circular with a diameter of 32 feet. What is the area of the playground? Round your answer to the nearest tenth.

803.8 ft²

14. Use a centimeter ruler to measure the radius of the circle. Then find the area of the shaded region of the circle. Use 3.14 for π. Round your answer to the nearest tenth.

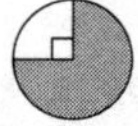

1.5 cm²

39

Holt Mathematics

The formula $A = \pi r^2$ is used to find the area of a circle. Since the value of π is about 3.14, you can use the formula $A \approx 3.14 \cdot r \cdot r$ to estimate the area of a circle. Remember that area is expressed in square units.

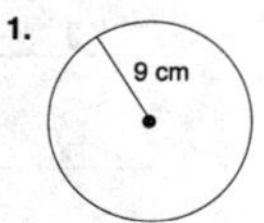
4 in.

The radius of the circle is 4 in.

$A \approx 3.14 \cdot r \cdot r$

$A \approx 3.14 \cdot 4 \cdot 4$

$A \approx 50.24$

The area of the circle is 50.2 in² to the nearest tenth.

Find the area of each circle to the nearest tenth. Use 3.14 for π.

1.
9 cm

The radius is ___9___ cm.

$A = \pi r^2$

$A \approx 3.14 \cdot$ ___9___ $\cdot$ ___9___

$A \approx$ ___254.34___

The area is ___254.3___ cm² to the nearest tenth.

2.
10 mm

The diameter is 10 mm.

The radius is ___5___ mm.

$A = \pi r^2$

$A \approx 3.14 \cdot$ ___5___ $\cdot$ ___5___

$A \approx$ ___78.5___

The area is ___78.5___ m² to the nearest tenth.

3. 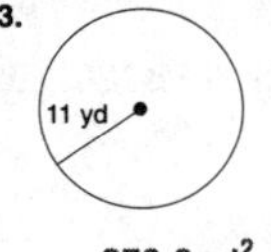
11 yd

379.9 yd²

4.
3 m

28.3 m²

5.
12 ft

113.0 ft²

6. What is the area of a circle with radius 13 yd? Round your answer to the nearest tenth.

530.7 yd²

40

Holt Mathematics

You can add or subtract to find the area of a shaded region.

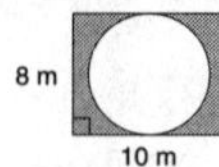
8 m
10 m

8 m
10 m

4 m

Find the area of the rectangle.
$A = \ell w = 10 \cdot 8 = 80$ m²

Find the area of the circle.
$A = \pi r^2 = \pi \cdot 4^2 \approx 50.24$ m²

The shaded area ≈ 80 m² $- 50.24$ m² ≈ 29.76 m² ≈ 29.8 m².

Add or subtract to find the area of the shaded region. Round your answer to the nearest tenth.

1.
5 yd

about 5.4 yd²

2.
4 ft
7.5 ft

about 17.4 ft²

3.
10 in.
7.1 in.

about 28.1 in²

4.
15 cm
25 cm
20 cm

about 340.6 cm²

5.
18 mm

about 69.7 mm²

6.
6 in.   6 in.

about 56.5 in²

7.
8 m

about 27.5 m²

8.
12 ft
12 ft
6 ft

about 31.0 ft²

9.
5 in.
3 in.

about 50.2 in²

41

Holt Mathematics

Write the correct answer.

1. According to the Royal Canadian Mint Act, a 50-cent Canadian coin must have a diameter of 27.13 millimeters. What is the area of this coin to the nearest tenth of a square millimeter?

577.8 mm²

2. By regulation, the diameter of a 25-cent Canadian coin is 23.88 millimeters. What is the area of this coin to the nearest tenth of a square millimeter?

447.6 mm²

3. There is a water reservoir beneath a circular garden to supply a fountain in the garden. The reservoir has a 26-inch diameter. The garden has a 12-foot diameter. How much of the garden does not contain the water reservoir?

15,747.1 in²

4. A frying pan has a diameter of 11 inches. What is the area to the nearest square inch of the smallest cover that will fit on top of the frying pan?

95 in²

Choose the letter of the correct answer.

5. In the state of Texas, Austin is about 80 miles northeast of San Antonio. What area is represented by all of the land within 80 miles of San Antonio?

A 251.2 mi²   **C** 5,024 mi²

B 502.4 mi²   **(D)** 20,096 mi²

6. A standard CD has a diameter of 12 centimeters. What is the area of a circular case that can be used to store a CD?

(F) 114 cm²   **H** 105 cm²

G 112 cm²   **J** 92 cm²

7. A round dining table has a diameter of 2.5 meters. A round tablecloth has a diameter of 3.5 meters. What is the area to the nearest tenth of a meter of the part of the tablecloth that will hang down the side of the table?

A 18.8 m²   **(C)** 4.7 m²

B 6.3 m²   **D** 1.0 m²

8. Justin just got his driver's license. His parents are giving him permission to drive within a 25-mile radius of his home. What is the area to which Justin is restricted when driving?

F 7,850 mi²   **H** 157 mi²

(G) 1,962.5 mi²   **J** 314 mi²

42

Holt Mathematics

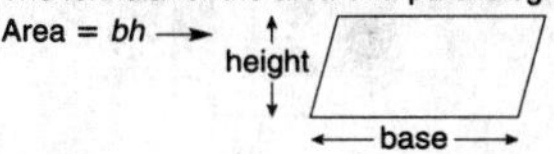
Reading Strategies
9-5 Use Graphic Aid

You can use what you know about the area of parallelograms to understand the area of circles.

The formula for the area of a parallelogram is:

Area = bh →

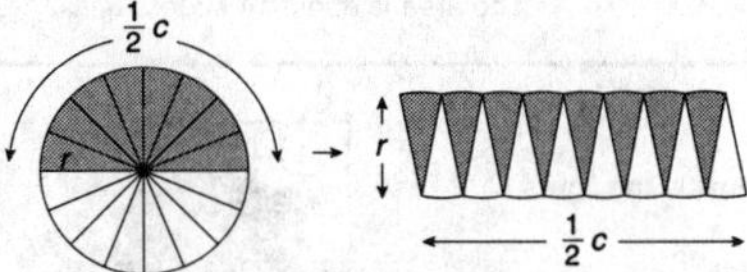
height — base

You can cut a circle into wedges and place the wedges next to each other to approximate a parallelogram.

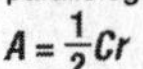

Compare the parallelogram to the circle to answer each question.

1. What figure do the wedges of a circle approximately form?

   __a parallelogram__

2. What is the formula for finding the area of a parallelogram?

   __$A = bh$__

3. A parallelogram has a measurement called the *base*. When a circle is reshaped into a parallelogram, what measurement is the same as the base?

   __half the circumference__

4. The other measurement of a parallelogram is the height. When a circle is reshaped into a parallelogram, what measurement is the same as the height?

   __the radius__

5. Use the information from Exercises 3 and 4 and the rearranged parallelogram to write the formula for the area of a circle.

   __$A = \frac{1}{2}Cr$__

43
Holt Mathematics

Puzzles, Twisters & Teasers
9-5 Area of Circles

Find the area of each circle and write it on the line. Use 3.14 for π. Then solve the riddle.

1. __28.26 m²__ = E
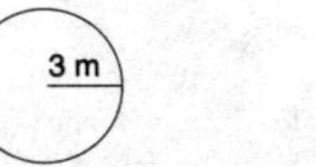
2. __50.24 in²__ = T
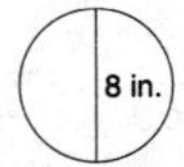
3. __78.5 in²__ = S
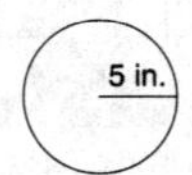

4. __200.96 cm²__ = L
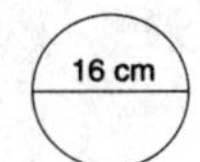
5. __314 yd²__ = U
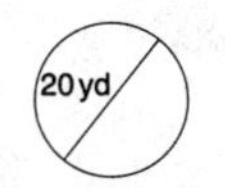
6. __32.15 yd²__ = I
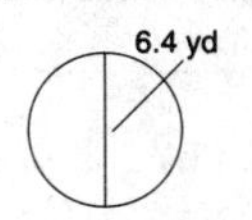

7. __63.58 ft²__ = O

8. __176.62 cm²__ = G
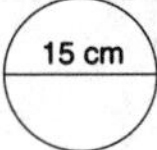
9. __3.8 m²__ = H
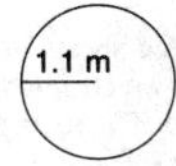

Which kind of house weighs the least?

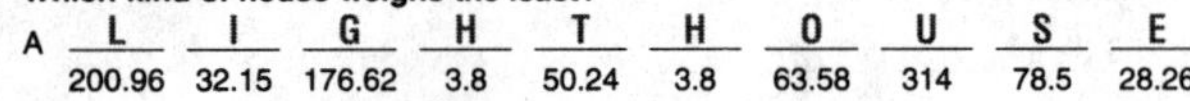

A __L__ __I__ __G__ __H__ __T__ __H__ __O__ __U__ __S__ __E__
  200.96 32.15 176.62 3.8 50.24 3.8 63.58 314 78.5 28.26

44
Holt Mathematics

Practice A
9-6 Area of Irregular Figures

Estimate the area of each figure. Each square represents 1 square foot. Choose the letter for the best answer.

1. 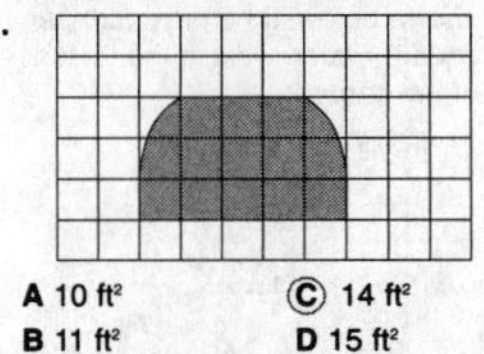

   A 10 ft²   C 14 ft²
   B 11 ft²   D 15 ft²

2. 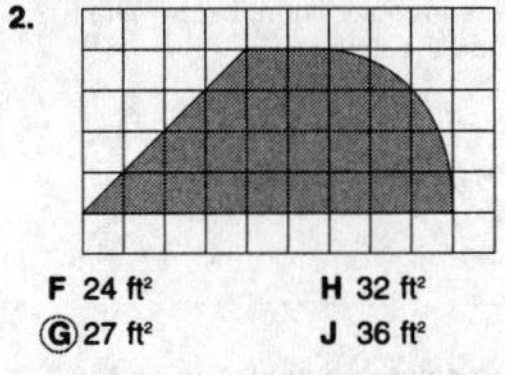

   F 24 ft²   H 32 ft²
   G 27 ft²   J 36 ft²

Find the area of each figure. Use 3.14 for π.

3. 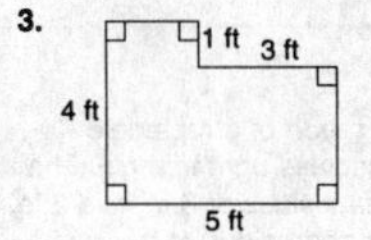

   __17 ft²__

4. 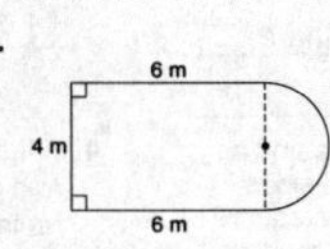

   __30.28 m²__

5. 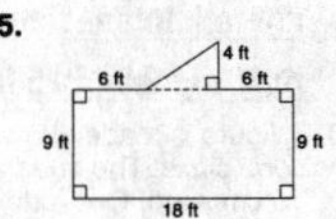

   __174 ft²__

6. 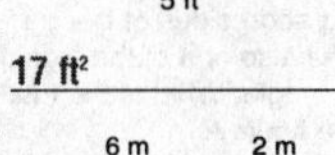

   __84 m²__

7. 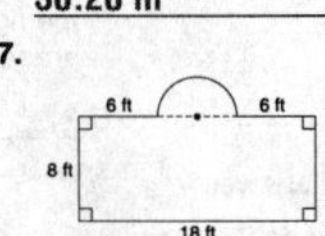

   __158.13 ft²__

8. 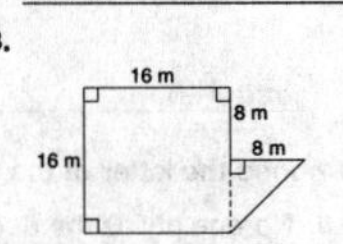

   __288 m²__

9. The figure shows the dimensions of a room. How much carpet is needed to cover the floor?

   __189.25 ft²__

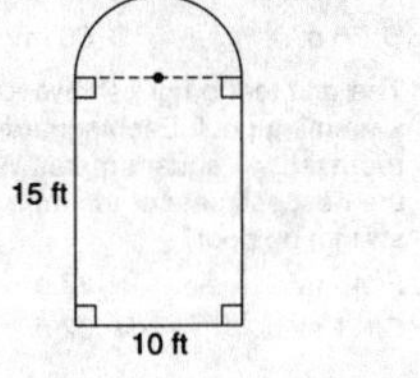

45
Holt Mathematics

Practice B
9-6 Area of Irregular Figures

Estimate the area of each figure. Each square represents 1 square foot.

1. 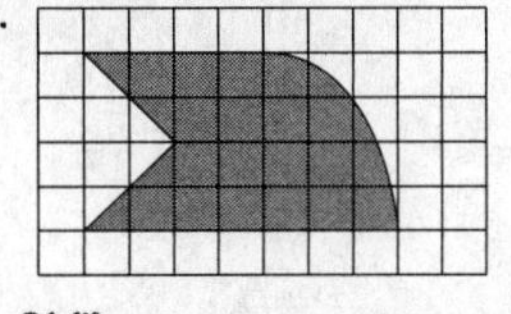

   __21 ft²__

2. 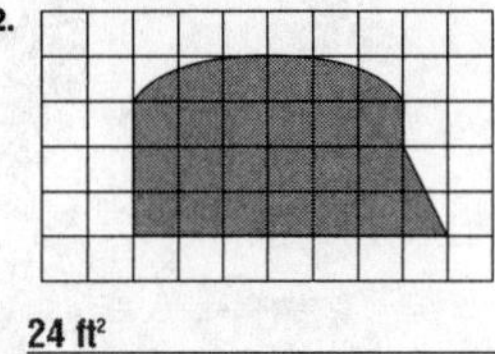

   __24 ft²__

Find the area of each figure. Use 3.14 for π.

3. 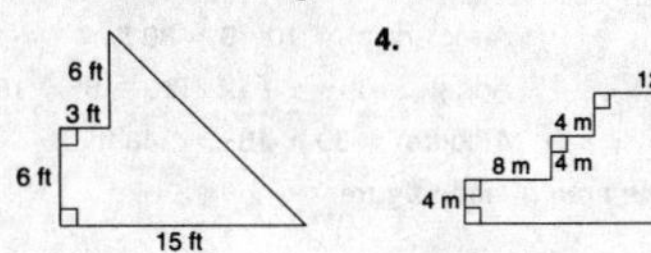

   __90 ft²__

4. 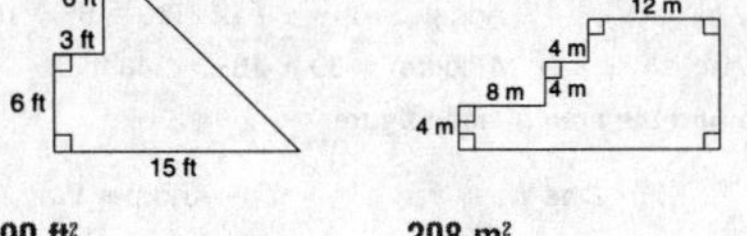

   __208 m²__

5.

   __140 ft²__

6.

   __23.13 m²__

7. 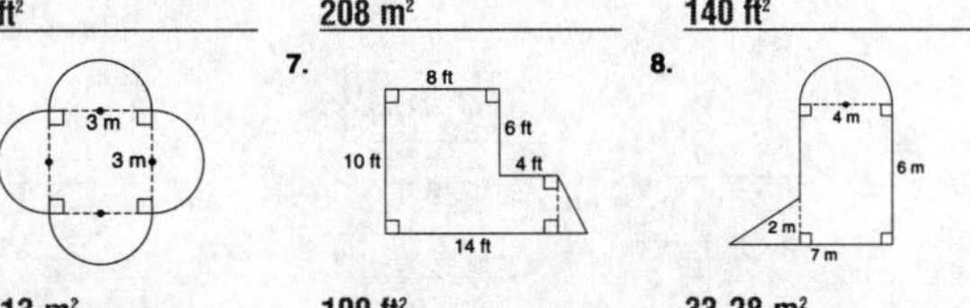

   __100 ft²__

8.

   __33.28 m²__

9. Marci is going to use tile to cover her terrace. How much tile does she need?

   __57.12 m²__

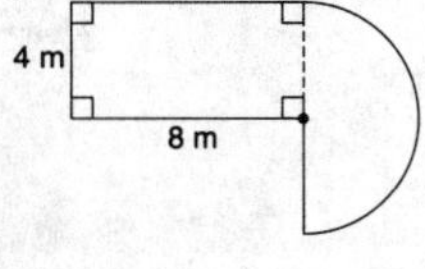

46
Holt Mathematics

Holt Mathematics

Estimate the area of each figure. Each square represents 1 square foot.

1. 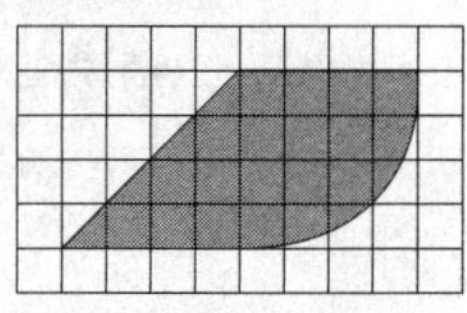

22 ft²

2. 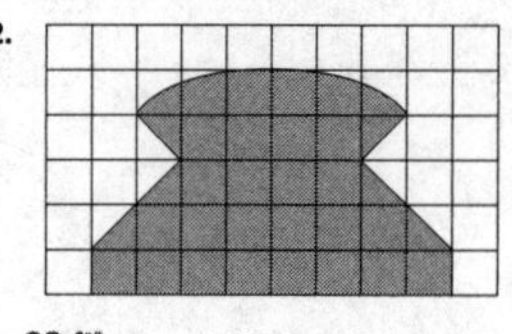

30 ft²

Find the area of each figure. Use 3.14 for π.

3. 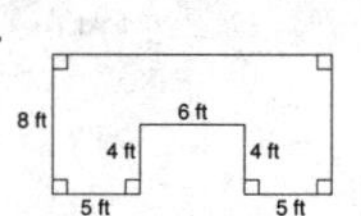

104 ft²

4. 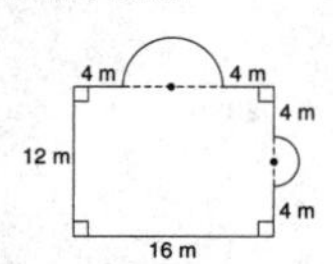

223.4 m²

5. 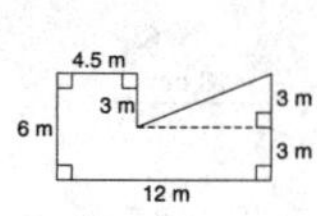

60.75 m²

6. The figure shows the dimensions of a room in which wedding receptions are held. The room is being carpeted. The three semi-circular parts of the room are congruent. How much carpet is needed?

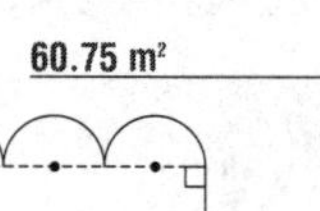

258.39 m²

7. A polygon has vertices at $F(-5, 2)$, $G(-3, 2)$, $H(-3, 4)$, $J(1, 4)$, $K(1, 1)$, $L(4, 1)$, $M(4, -2)$, $N(6, -2)$, $P(6, -3)$, and $Q(5, -3)$. Graph the figure on the coordinate plane. Then find the area and the perimeter of the figure.

A = 52 units²; P = 36 units

47

When an irregular figure is on graph paper, you can estimate its area by counting whole squares and parts of squares. Follow these steps.

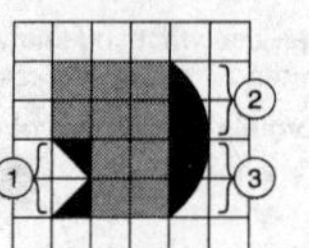

- Count the number of whole squares. There are 10 whole squares.
- Combine parts of squares to make whole squares or $\frac{1}{2}$-squares

  Section 1 = 1 square
  Section 2 ≈ $1\frac{1}{2}$ squares
  Section 3 ≈ $1\frac{1}{2}$ squares

- Add the whole and partial squares.

  $10 + 1 + 1 + 1\frac{1}{2} + 1\frac{1}{2} = 14$

  The area is about 14 square units.

Estimate the area of the figure.

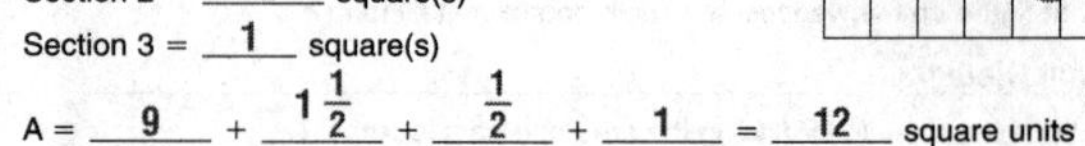

1. There are ___9___ whole squares in the figure.

Section 1 ≈ $1\frac{1}{2}$ square(s)

Section 2 = $\frac{1}{2}$ square(s)

Section 3 = 1 square(s)

$A = \underline{9} + \underline{1\frac{1}{2}} + \underline{\frac{1}{2}} + \underline{1} = \underline{12}$ square units

You can break an irregular figure into shapes that you know. Then use those shapes to find the area.

A (rectangle) $= 9 \cdot 6 = 54$ m²
A (square) $= 3 \cdot 3 = 9$ m²
A (irregular figure) $= 54 + 9 = 63$ m²

Find the area of the figure.

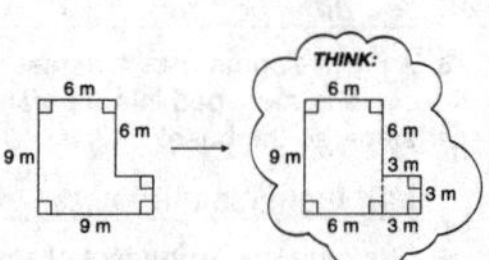

2. A (rectangle) $= \underline{32}$ ft²

A (triangle) $= \underline{6}$ ft²

A (irregular figure) $= \underline{32} + \underline{6}$

$= \underline{38}$ ft²

48

Sometimes there is more than one way to find the area of an irregular figure.

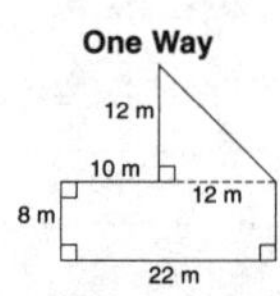

One Way

A(rectangle) $= 22 \cdot 8 = 176$ ft²
A(triangle) $= \frac{1}{2} \cdot 12 \cdot 12 = 72$ ft²
A(figure) $= 176 + 72 = 248$ ft²

Another Way

A(rectangle) $= 10 \cdot 8 = 80$ ft²
A(trapezoid) $= \frac{1}{2} \cdot 12 \cdot (20 + 8) = 168$ ft²
A(figure) $= 80 + 168 = 248$ ft²

Show two different ways to find the area of each figure.

1. 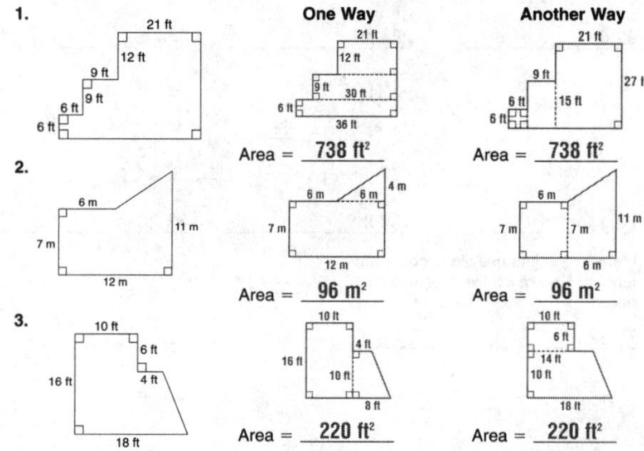

One Way Area = ___738 ft²___

Another Way Area = ___738 ft²___

2.

One Way Area = ___96 m²___

Another Way Area = ___96 m²___

3.

One Way Area = ___220 ft²___

Another Way Area = ___220 ft²___

49

Write the correct answer.

1. Explain how to find the area of the irregular figure below. Then find the area.

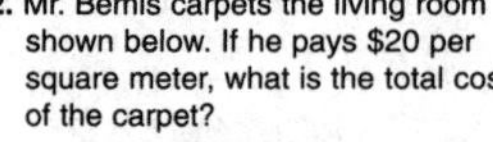

Possible answer: Divide the figure into a rectangle and a right triangle; 325 ft²

2. Mr. Bemis carpets the living room shown below. If he pays $20 per square meter, what is the total cost of the carpet?

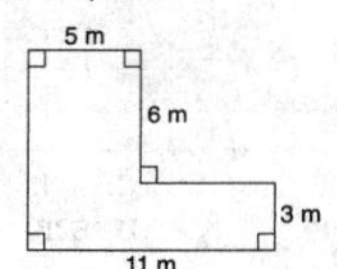

$1,260

3. A figure is made of a square and a semi-circle. The square has sides of 16 cm each. One side of the square is also the diameter of the semi-circle. What is the total area of the figure?

356.48 cm²

4. A figure is made of a rectangle and an isosceles right triangle. The rectangle has sides of 6 in. and 3 in. One of the short sides of the rectangle is also one of the legs of the right triangle. What is the total area of the figure?

22.5 in.²

Choose the letter of the correct answer.

5. Norene builds the deck at the right. The area of the deck is 10 m² greater than was originally planned. What is the area of the deck?

A 110 m²
B 76 m²
C 66 m²
D 56 m²

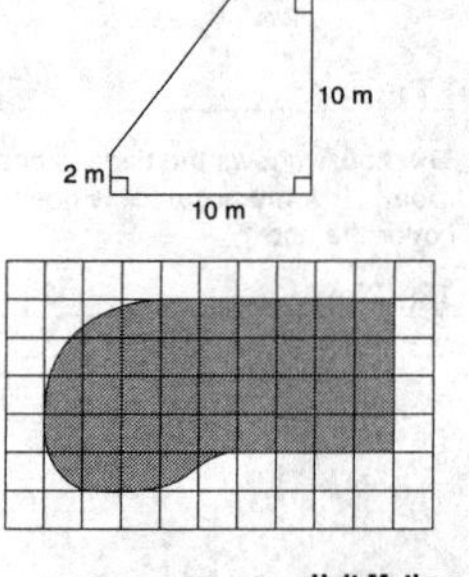

6. The grid to the right shows a swimming pool. Each square represents 1 square meter. What is the best estimate of the area of the swimming pool?

F 45 m²
G 41 m²
H 37 m²
J 32 m²

50

80

Sometimes an irregular figure is made of shapes that you know, such as rectangles, triangles, or semi-circles.

You can draw a line to break an irregular figure into other shapes. Then you can use those shapes to help you find the area.

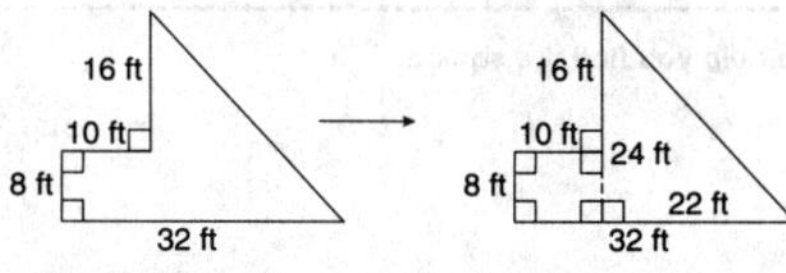

Answer each question.

1. A dashed line has been drawn through the figure. What two shapes does the dashed line create?

   a rectangle and a right triangle

2. How can you find the area of the rectangle?

   Use the formula, area = length · width.

3. What is the area of the rectangle?

   80 ft²

4. How can you find the area of the right triangle?

   Use the formula, area = $\frac{1}{2}$ · base · height.

5. What is the area of the right triangle?

   264 ft²

6. How can you use the area of the rectangle and the area of the right triangle to find the total area of the irregular figure?

   Add the area of the rectangle to the area of the right triangle

7. What is the total area of the irregular figure?

   344 ft²

 Holt Mathematics

Each square represents 1 square foot. Estimate the area of each figure. Circle the letter next to the better estimate.

1. 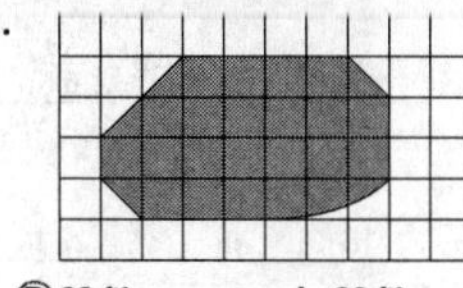

2. 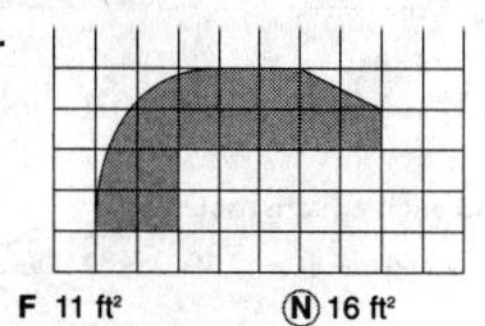

   (D) 23 ft²   I 26 ft²   F 11 ft²   (N) 16 ft²

Next, find the area of each figure. Circle the letter next to your answer.

3.

   L 38 m²
   (R) 42 m²

4. 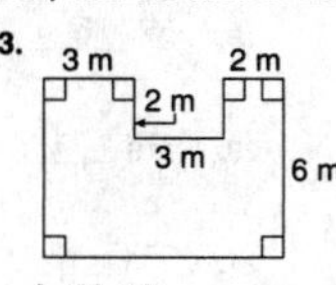

   (A) 54.28 ft²
   E 60.56 ft²

5. 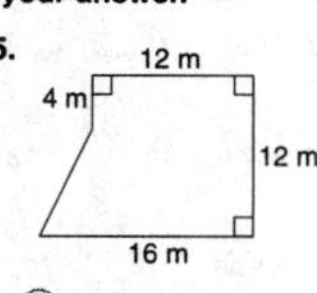

   (O) 160 m²
   E 168 m²

6.

   H 52 m²
   (L) 60 m²

7.

   S 89.12 ft²
   (Y) 114.24 ft²

8. 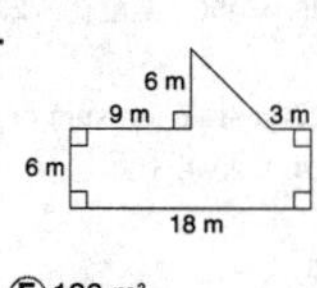

   (F) 126 m²
   B 144 m²

Write the circled letters above the problem numbers to solve the riddle.

What kind of insect breathes fire?

A D R A G O N F L Y
  1. 3. 4.   5. 2. 8. 6. 7.

 Holt Mathematics

Find each square.

1. 5² __25__   2. 10² __100__   3. 11² __121__

4. 13² __169__   5. 7² __49__   6. 12² __144__

Find each square root.

7. √225 __15__   8. √36 __6__   9. √81 __9__

10. √289 __17__   11. √196 __14__   12. √484 __22__

Match each square root in Column A with the letter of its nearest whole number estimate in Column B. Use a calculator to check your answer.

Column A	Column B	Column A	Column B
13. √38 __C__	A. 2	19. √84 __K__	G. 10
14. √19 __F__	B. 8	20. √46 __J__	H. 4
15. √66 __B__	C. 6	21. √95 __G__	I. 8
16. √54 __E__	D. 5	22. √10 __L__	J. 7
17. √29 __D__	E. 7	23. √15 __H__	K. 9
18. √5 __A__	F. 4	24. √71 __I__	L. 3

25. The area of a checkerboard is 112 in². What is the approximate length of each side of the checkerboard? Find your answer to the nearest inch.

   11 in.

 Holt Mathematics

Find each square.

1. 6² __36__   2. 19² __361__   3. 13² __169__   4. 4² __16__

5. 10² __100__   6. 14² __196__   7. 20² __400__   8. 18² __324__

Find each square root.

9. √289 __17__   10. √49 __7__   11. √256 __16__   12. √81 __9__

13. √121 __11__   14. √625 __25__   15. √576 __24__   16. √900 __30__

17. √11 __3__   18. √31 __6__   19. √98 __10__   20. √50 __7__

21. √152 __12__   22. √14 __4__   23. √70 __8__   24. √28 __5__

25. √39 __6__   26. √193 __14__   27. √119 __11__   28. √85 __9__

29. √5 __2__   30. √42 __6__   31. √75 __9__   32. √215 __15__

33. The area of a square vegetable garden is 75 ft². What is the approximate length of each side of the garden? Find your answer to the nearest foot.   **9 ft**

34. The area of a computer screen is 138 in². What is the approximate length of each side of the screen? Find your answer to the nearest inch.   **12 in.**

35. Tim broke a square picture window with his baseball. The area of the window is 52 ft². What is the approximate width of the window to be replaced? Find your answer to the nearest foot.   **7 ft**

36. A square tile has an area of 413 cm². What is the approximate length of a side of the tile? Find your answer to the nearest centimeter.   **20 cm**

 Holt Mathematics

 Holt Mathematics

Practice C
Powers and Roots

Find each square

1. 18^2 __324__   2. $(0.7)^2$ __0.49__   3. 27^2 __729__   4. $(3.9)^2$ __15.21__

5. 50^2 __2,500__   6. $(0.31)^2$ __0.0961__   7. $\left(\frac{1}{5}\right)^2$ __$\frac{1}{25}$__   8. $\left(\frac{7}{8}\right)^2$ __$\frac{49}{64}$__

Find each square root.

9. $\sqrt{784}$ __28__   10. $\sqrt{484}$ __22__   11. $\sqrt{225}$ __15__   12. $\sqrt{961}$ __31__

Estimate each square root to the nearest whole number. Use a calculator to check your answer.

13. $\sqrt{54}$ __7__   14. $\sqrt{86}$ __9__   15. $\sqrt{12}$ __3__   16. $\sqrt{27}$ __5__

17. $\sqrt{71}$ __8__   18. $\sqrt{15}$ __4__   19. $\sqrt{95}$ __10__   20. $\sqrt{118}$ __11__

21. $\sqrt{149}$ __12__   22. $\sqrt{201}$ __14__   23. $\sqrt{250}$ __16__   24. $\sqrt{175}$ __13__

25. $\sqrt{450}$ __21__   26. $\sqrt{375}$ __19__   27. $\sqrt{600}$ __24__   28. $\sqrt{1,000}$ __32__

Estimate each sum or difference to the nearest whole number.

29. $\sqrt{20} + \sqrt{30}$ __9__   30. $\sqrt{50} + \sqrt{10}$ __10__   31. $\sqrt{75} - \sqrt{60}$ __1__

32. $\sqrt{7} + 7^2$ __52__   33. $5^2 - \sqrt{5}$ __23__   34. $\sqrt{13} + 13^2$ __173__

35. Find the area of a square whose perimeter is 48 in.   __144 in^2__

36. Find the perimeter of a square whose area is 225 yd^2.   __60 yd__

37. The area of a square rug is 3,000 in^2. What is the approximate length of each side of the rug? Find your answer to the nearest inch.   __55 in.__

Holt Mathematics

Reteach
Powers and Roots

Recall that the formula for the area of a square is $A = s^2$. A power in which the exponent is 2 is called a *square*.

Use the model to help you find the square.

1. 5^2

$A = s^2$
$A = $ __5__ ^{2}
$A = $ __5__ $\cdot$ __5__
$A = $ __25__

2. 9^2

$A = s^2$
$A = ($ __9__ $)^2$
$A = $ __9__ $\cdot$ __9__
$A = $ __81__

Find the square.

3. 13^2 __169__   4. 16^2 __256__   5. 11^2 __121__   6. 7^2 __49__

The numbers 36 and 81 are *perfect squares*. **Perfect squares** are numbers that are the squares of whole numbers.

The square root of 36 is 6. $\sqrt{36} = 6$ because $6 \cdot 6 = 36$.

A table of square roots can help you estimate the square root of a number that is not a perfect square.

Square Root	1	2	3	4	5	6	7	8	9	10
Perfect Squares	1	4	9	16	25	36	49	64	81	100

7. $\sqrt{44}$
Find the perfect square nearest 44.
44 is closest to __49__.
Since $\sqrt{49} = $ __7__,
$\sqrt{44}$ is closest to __7__.

8. $\sqrt{87}$
Find the perfect square nearest 87.
87 is closest to __81__.
Since $\sqrt{81} = $ __9__,
$\sqrt{87}$ is closest to __9__

Holt Mathematics

Challenge
Power Patterns

1. Complete the tables below.

Powers of 3	
3^1	3
3^2	9
3^3	27
3^4	81
3^5	243
3^6	729
3^7	2,187
3^8	6,561

Powers of 7	
7^1	7
7^2	49
7^3	343
7^4	2,401
7^5	16,807
7^6	117,649
7^7	823,543
7^8	5,764,801

Powers of 9	
9^1	9
9^2	81
9^3	729
9^4	6,561
9^5	59,049
9^6	531,441
9^7	4,782,969
9^8	43,046,721

2. Describe the pattern for the ones digits of the powers of 3 values.
__3, 9, 7, 1 repeating__

3. What is the ones digit of 3^9? of 3^{31}? of 3^{98}?   __3; 7; 9__

4. Describe the pattern of the ones digits of the powers of 7 values.
__7, 9, 3, 1 repeating__

5. What is the ones digit of 7^{10}? of 7^{24}? of 7^{53}?   __9; 1; 7__

6. Do the ones digits for the values of the powers of all whole numbers follow a pattern of *four* repeating digits? Explain.
__No; The powers of 9 have only two repeating digits.__

7. What is the ones digit of 11^3? of 11^4? of 11^5? of 11^6?   __1; 1; 1; 1__

8. The ones digits of the values of the powers of 11 are the same as the ones digits of the powers of what number?   __1__

9. Describe the patterns shown by the powers of 5.
__Except for 5^1, the last 2 digits always form 25.__

10. Describe the pattern of the ones digits for the powers of 6.
__They are all 6's.__

Holt Mathematics

Problem Solving
Powers and Roots

Write the correct answer. For Problems 1 and 2, use the following formula to find the distance in meters a free-falling object falls from a place of rest: $d = 0.5 \cdot 9.8 \cdot t^2$ (t = time in seconds).

1. As part of a science experiment, Hsing drops a ball from the roof of the school. How far does the ball fall in 2 seconds?
__19.6 m__

2. Mel drops a stone from the edge of a cliff overlooking the ocean. How far does the stone fall in 5 seconds?
__122.5 m__

3. At the county fair, the apple pies are lined up side-by-side for judging on a 6-foot table. Each pie has an area of 50.24 in^2. How many pies are on the table?
__9 pies__

4. The community swimming pool has an area of 1,024 square feet. The pool is in the shape of a square. What is the perimeter of the pool?
__128 ft__

Choose the letter of the correct answer.

5. The Portuguese national flag is a rectangle. In the center of the flag is a coat of arms and shield on a circle. This circle has a diameter that is half the flag's height. If the flag's circle has an area of 3.14 square feet, what is the height of the flag?
A 1 ft
B 3 ft
Ⓒ 4 ft
D 2 ft

6. A square picture has an area of 81 square inches. The perimeter of the frame for the picture is 8 inches longer than the perimeter of the picture itself. What is the length of each side of the square frame for this picture?
F 8 in.
G 10 in.
H 9 in.
Ⓙ 11 in.

7. A basketball game starts with a jump ball. This occurs in the center of the basketball court within a circle that has an area of 113.04 square feet. What is the radius of this circle?
A 3 ft
Ⓑ 6 ft
C 9 ft
D 36 ft

8. A square fence encloses a vegetable garden with an area of 169 square feet. What is the perimeter of the fence?
Ⓕ 52 ft
G 26 ft
H 13 ft
J 56.25 ft

Holt Mathematics

Holt Mathematics

Reading Strategies
Use A Graphic Organizer

A **square number** is a whole number multiplied by itself.

$3 \times 3 = 9$ 9 is a square number.

$6 \times 6 = 36$ 36 is a square number.

A square number has two factors that are the same.

A **square root** is the opposite of a square number.

This chart helps you picture square numbers and square roots.

Perfect Square	Not a Perfect Square
4 by 4 $A = \ell w \ (4 \cdot 4)$ $A = 4^2 \rightarrow$ Read: "four squared." $A = 16$ square units 16 is a square number.	4 by 5 $A = \ell w \ (4 \cdot 5)$ $A = 20$ $A = 20$ square units 20 is not a square number.

Squares and Square Roots

Square Root

- The opposite of a square
- $\sqrt{}$ is the radical symbol used to write square roots.

$\sqrt{16} = 4$ $\sqrt{20} \approx 4.47$

Read: "The square root of 16 equals 4."   Read: "The square root of 20 equals approximately 4.47."

Use the chart to help you answer each question.

1. How can you tell if a number is a square number?

   It has two factors that are the same.

2. Write how you would read 5^2.

   five squared

3. Complete. $5^2 = $ __________ 25

4. Write how you would read. $\sqrt{25}$.

   the square root of 25

Puzzles, Twisters & Teasers
The Root of the Matter!

Find and circle the words in the box below in the word search. The words may be horizontal or vertical. Find a word in the word search that solves the riddle. Circle it and write it on the line.

power	root	perfect	square	radical
sign	express	value	symbol	model

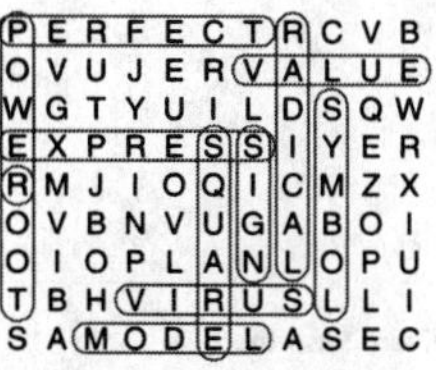

Why did the computer sneeze?

It had a ______ virus ______ .

Practice A
The Pythagorean Theorem

Use the Pythagorean Theorem to find each missing measure. Choose the letter for the best answer.

1. 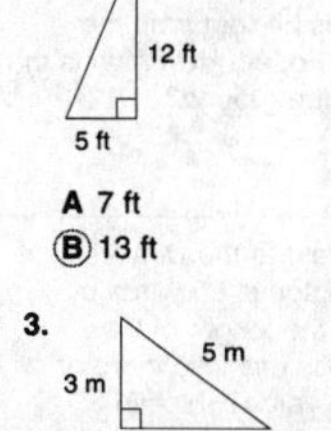

   A 7 ft   C 14 ft
   Ⓑ 13 ft   D 17 ft

2. 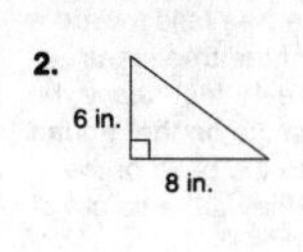

   F 14 in.   H 5 in.
   Ⓖ 10 in.   J 2 in.

3. 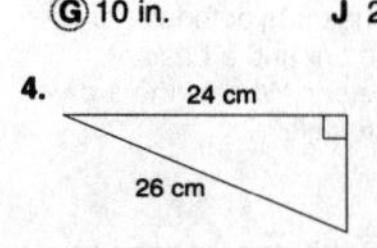

   A 2 m   C 6 m
   Ⓑ 4 m   D 8 m

4.

   F 2 cm   Ⓗ 10 cm
   G 4 cm   J 40 cm

5. 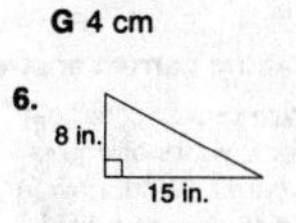

   A 11 yd   Ⓒ 20 yd
   B 16 yd   D 28 yd

6.

   F 23 in.   H 13 in.
   Ⓖ 17 in.   J 7 in.

7. A 15-foot ladder is leaning against a wall. The ladder is 6 ft from the base of the wall. About how far above the ground does the ladder touch the wall? Round your answer to the nearest tenth.

   about 13.7 ft

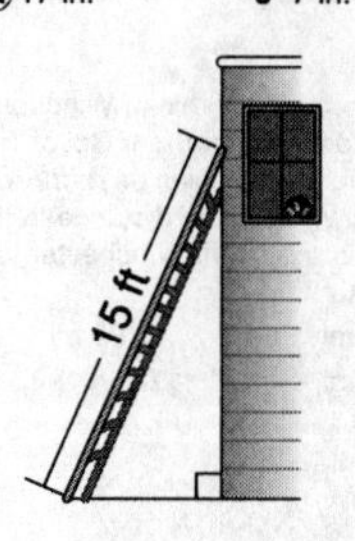

Practice B
The Pythagorean Theorem

Use the Pythagorean Theorem to find each missing measure.

1.

   15 in.

2.

   8 cm

3.

   26 ft

4.

   24 m

5.

   5 yd

6. 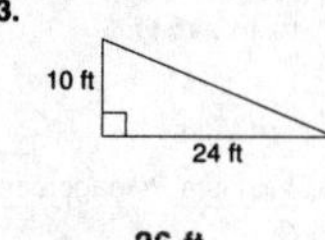

   10 m

7. 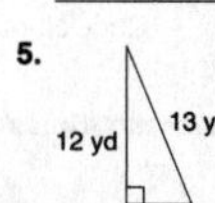

   16 mm

8.

   30 in.

9. 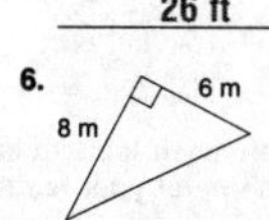

   15 yd

10. A 20-ft ladder is leaning against a wall. If the ladder is 12 ft from the base of the wall, how high above the ground does the ladder touch the wall?

    16 ft

11. A checkerboard is 10 inches long on each side. What is the length of the diagonal from one corner to another? Round your answer to the nearest tenth.

    14.1 in.

12. Chang lives 8 miles east of the school. Deborah lives 12 miles south of the school. Approximately how far apart are Chang's and Deborah's homes? Round your answer to the nearest tenth.

    14.4 mi

13. A rectangle is 26 meters long and 18 meters wide. What is the length of the diagonal of the rectangle to the nearest meter?

    32 m

Practice C
The Pythagorean Theorem

Use the Pythagorean Theorem to find each missing measure.

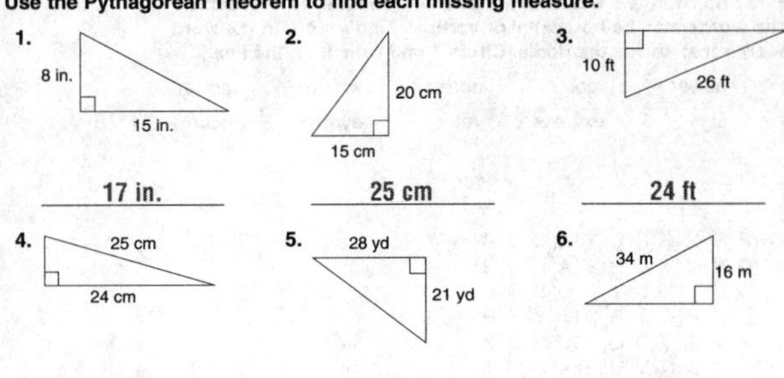

1. 8 in. / 15 in.

17 in.

2. 20 cm / 15 cm

25 cm

3. 10 ft / 26 ft

24 ft

4. 25 cm / 24 cm

7 cm

5. 28 yd / 21 yd

35 yd

6. 34 m / 16 m

30 m

The lengths of two sides of a right triangle are given. Find the length of the third side to the nearest tenth.

7. legs: 4 ft and 9 ft

9.8 ft

8. leg: 8 m; hypotenuse: 12 m

8.9 m

9. leg: 15 ft; hypotenuse: 20 ft

13.2 ft

10. legs: 10 m and 7 m

12.2 m

Determine whether each set of lengths forms a right triangle.

11. $a = 10$, $b = 24$, $c = 25$ **no**

12. $a = 12$, $b = 16$, $c = 20$ **yes**

13. $a = 4$, $b = 6$, $c = 8$ **no**

14. $a = 5$, $b = 12$, $c = 13$ **yes**

Solve.

15. The foot of a ladder is 8 ft from the base of a wall. If the ladder is 20 ft long, how high up on the wall does the ladder reach? Round your answer to the nearest tenth. **18.3 ft**

16. The hypotenuse of an isosceles right triangle is 16 centimeters. Find the length of the two other sides of the triangle to the nearest tenth. **11.3 cm**

Holt Mathematics

Reteach
The Pythagorean Theorem

A triangle containing a right angle is called a *right triangle*. The sides adjacent to the right angle are the **legs**, represented by a and b. The side opposite the right angle is the **hypotenuse**, represented by c.

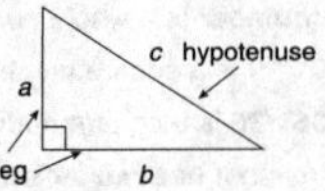

The **Pythagorean Theorem** states:

If a and b are the lengths of the legs of a right triangle and c is the length of the hypotenuse, then $a^2 + b^2 = c^2$.

Use the Pythagorean Theorem to find each missing length.

1. 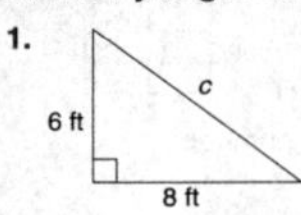 6 ft / c / 8 ft

$$a^2 + b^2 = c^2$$
$$\underline{6}^2 + \underline{8}^2 = c^2$$
$$\underline{36} + \underline{64} = c^2$$
$$\underline{100} = c^2$$
$$c = \underline{10 \text{ ft}}$$

2. 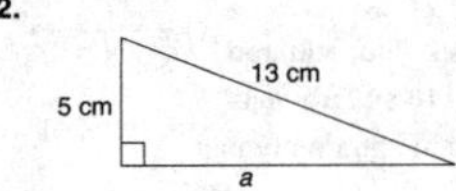 5 cm / 13 cm / a

$$a^2 + b^2 = c^2$$
$$a^2 + \underline{5}^2 = \underline{13}^2$$
$$a^2 + \underline{25} = \underline{169}$$
$$a^2 = \underline{144}$$
$$a = \underline{12 \text{ cm}}$$

3. 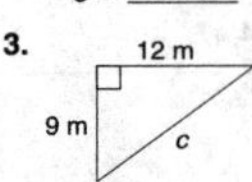 12 m / 9 m / c

15 m

4. 34 in. / a / 30 in.

16 in.

5. 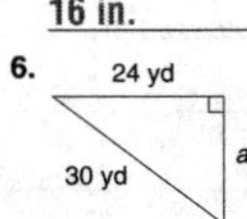 c / 16 mm / 12 mm

20 mm

6. 24 yd / 30 yd / a

18 yd

Holt Mathematics

Challenge
Triangle Families

Any three numbers that satisfy the formula $a^2 + b^2 = c^2$ form a *Pythagorean triple*.

$$3^2 + 4^2 = 5^2$$
$$9 + 16 = 25$$

(3, 4, 5) is a Pythagorean triple.

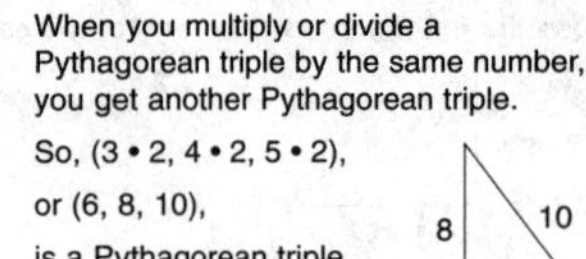 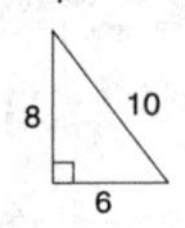

When you multiply or divide a Pythagorean triple by the same number, you get another Pythagorean triple.

So, $(3 \cdot 2, 4 \cdot 2, 5 \cdot 2)$,
or (6, 8, 10),
is a Pythagorean triple.

Check it out: $6^2 + 8^2 = 10^2$
$36 + 64 = 100$

Complete each table to find new Pythagorean triples. Then check some of your results.

1. **The (3, 4, 5) Family**

Multiply by 3.	(9, 12, 15)
Multiply by 5.	(15, 20, 25)
Multiply by 10.	(30, 40, 50)
Divide by 5.	$\left(\frac{3}{5}, \frac{4}{5}, 1\right)$

2. **The (5, 12, 13) Family**

Multiply by 2.	(10, 24, 26)
Multiply by 3.	(15, 36, 39)
Multiply by 8.	(40, 96, 104)
Divide by 10.	$\left(\frac{1}{2}, \frac{6}{5}, \frac{13}{10}\right)$

3. **The (7, 24, 25) Family**

Divide by 4.	$\left(\frac{7}{4}, 6, \frac{25}{4}\right)$
Multiply by 2.	(14, 48, 50)
Multiply by 4.	(28, 96, 100)
Multiply by 12.	(84, 288, 300)

4. **The (8, 15, 17) Family**

Divide by 2.	$\left(4, \frac{15}{2}, \frac{17}{2}\right)$
Multiply by 3.	(24, 45, 51)
Multiply by 4.	(32, 60, 68)
Multiply by 20.	(160, 300, 340)

5. Use the formula to check one triple of the (3, 4, 5) family.

Possible answer: $15^2 + 20^2 = 25^2$; $225 + 400 = 625$

6. Use the formula to check one triple of the (5, 12, 13) family.

Possible answer: $10^2 + 24^2 = 26^2$; $100 + 576 = 676$

7. Use the formula to check one triple of the (7, 24, 25) family.

Possible answer: $28^2 + 96^2 = 100^2$; $784 + 9,216 = 10,000$

Holt Mathematics

Problem Solving
The Pythagorean Theorem

Write the correct answer.

1. During a storm, a tree falls toward a house. The top of the tree leans against the house 45 feet above the ground. The distance on the ground from the house to the base of the tree is 24 feet. What is the height of the tree? **51 ft**

2. During a training exercise, a firefighter leans a 40-foot ladder up to a window in a house. The bottom of the ladder is 24 feet from the bottom of the house. How high is the window from the ground? **32 ft**

3. A triangle has a hypotenuse of 25 centimeters and a base of 20 centimeters. What is the area of this right triangle? **150 cm²**

4. The football field at the University of Texas at Arlington is 60 yards by 100 yards. Is the length of the diagonal across this field more or less than 200 yards? Explain. **less, because $\sqrt{13,600}$ is less than 200**

Choose the letter of the correct answer.

5. The minimum size of a soccer field for players under 8 years of age is 20 yards by 30 yards. About how far is the diagonal distance on a field with these dimensions?
A about 12 yd   **C about 36 yd**
B about 25 yd   D about 45 yd

6. The minimum size of a soccer field for international matches is 70 yards by 110 yards. If a player runs diagonally across this field, about how much farther does she run than if the field were 50 yards by 100 yards?
F about 242 yd   H about 112 yd
G about 130 yd   **J about 19 yd**

7. In the state of Virginia, Winchester is 21 miles north of Front Royal. Arlington is 58 miles east of Front Royal. What is the distance to the nearest mile from Winchester to Arlington?
A 37 mi   C 89 mi
B 62 mi   D 441 mi

8. On a child's slide, the distance from the bottom rung to the top of the ladder is 6 feet. The straight distance from the bottom rung of the ladder to the bottom of the slide is 36 inches. About how long is the slide?
F about 6.7 ft   H about 9.0 ft
G about 8.4 ft   J about 36.5 ft

Holt Mathematics

Holt Mathematics

Reading Strategies
Understanding Vocabulary

A **right triangle** has special features.

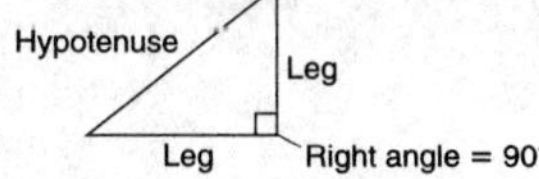

Use the triangle to complete each sentence.

1. A right triangle has one right angle that measures ______90______ degrees.

2. The two shorter sides of a right angle are called ______legs______.

3. The longest side of a right triangle is called the ______hypotenuse______.

The **Pythagorean Theorem** describes a special relationship among the sides of a right triangle. The theorem states that the sum of the legs squared is equal to the hypotenuse squared.

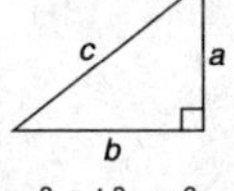

$a^2 + b^2 = c^2$

Look at the figure of the triangle to answer each question.

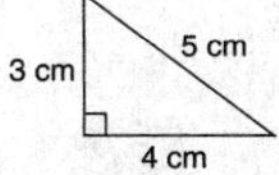

4. What is 3 cm squared? ________9 cm________

5. What is 4 cm squared? ________16 cm________

6. What is the total of the squares of the legs? ________25 cm________

7. What is 5 cm squared? ________25 cm________

8. What do you notice about the sum of the squares of the legs and the square of the hypotenuse? ________They are the same.________

67   **Holt Mathematics**

Puzzles, Twisters & Teasers
Fair and Square!

Use the Pythagorean Theorem to find the length of the hypotenuse of each triangle. Round your answers to the nearest tenth, if necessary. Write your answer on the line next to the letter. Replace the answers listed at the end with its corresponding letter. If you calculated the lengths correctly you will solve the riddle!

1. ____13 cm____ = K   2. ____15 m____ = E   3. ____20 m____ = N

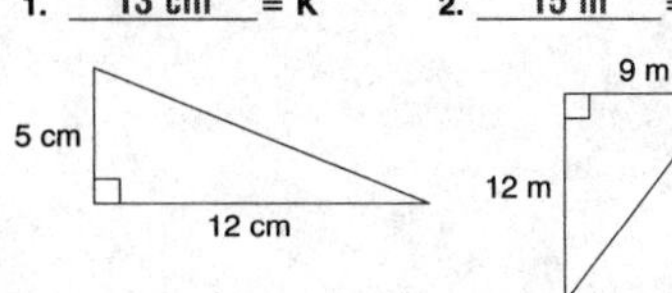

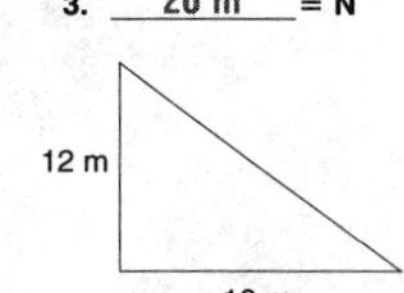

4. ____17 ft____ = B   5. ____5 cm____ = U   6. ____10 in.____ = I

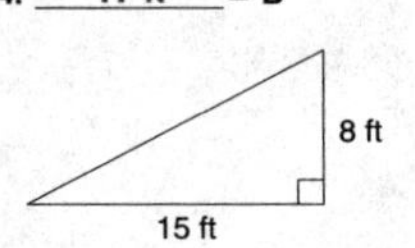

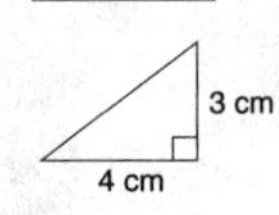

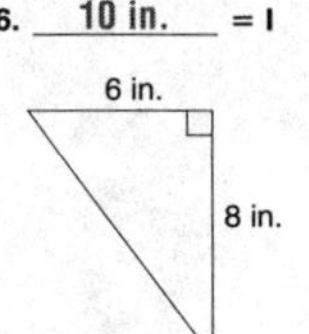

7. ____5.8 m____ = R   8. ____4.2 yd____ = G   9. ____127.3 cm____ = T

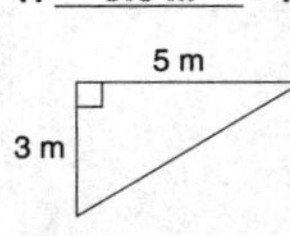

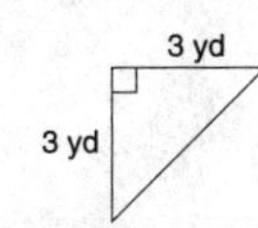

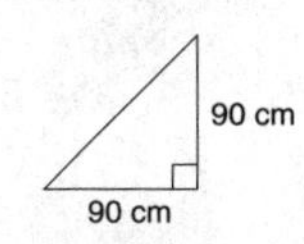

Why did the woman want an elephant instead of a car?

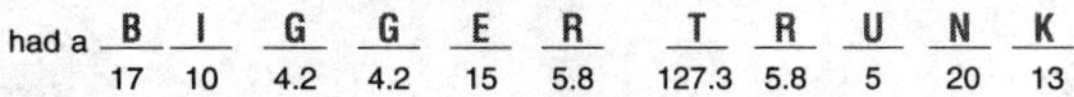

It had a B I G G E R T R U N K .
      17 10 4.2 4.2 15 5.8 127.3 5.8 5 20 13

68   **Holt Mathematics**

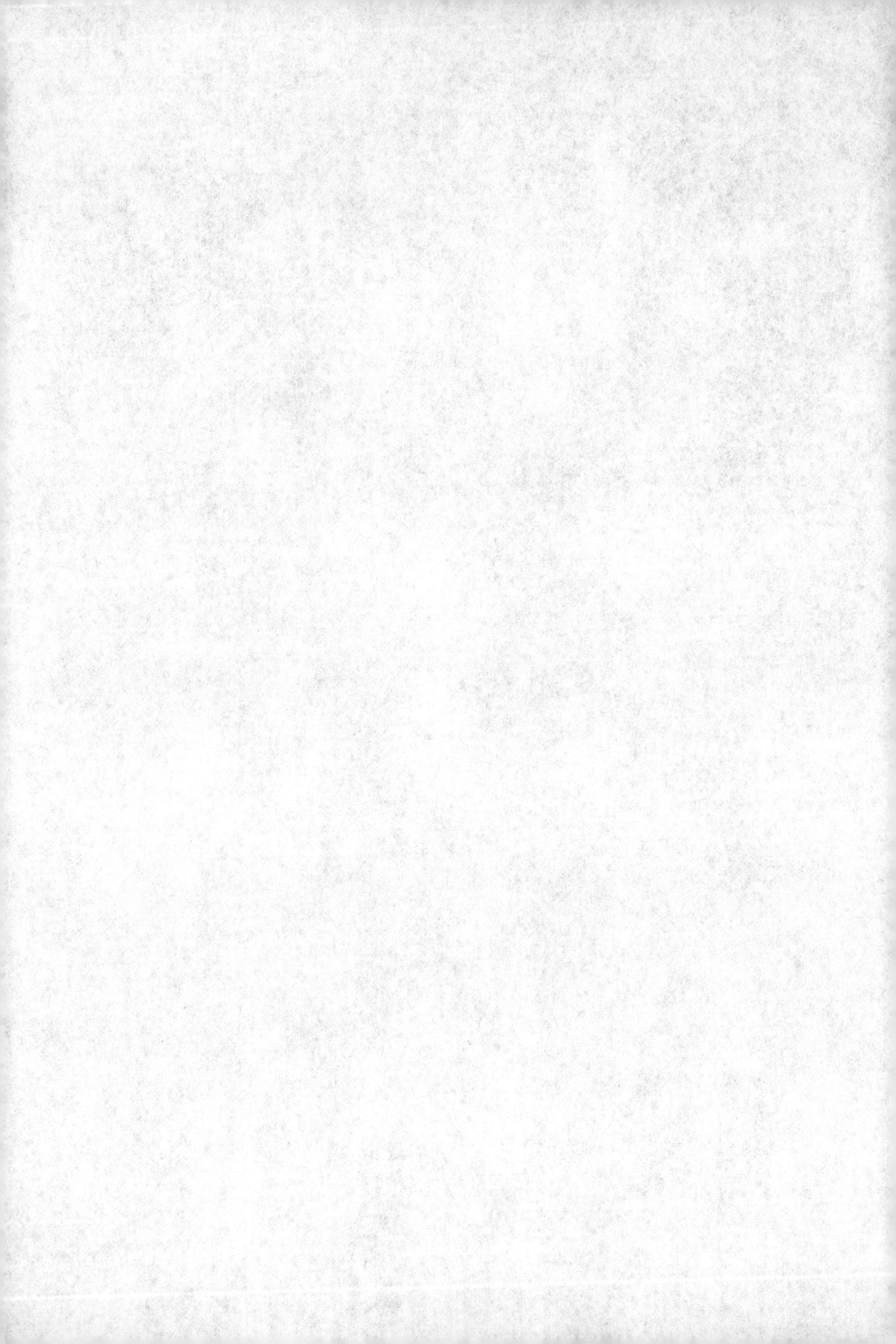